The Life Science

the text of this book is printed
on 100% recycled paper

The Life Science

Current Ideas of Biology

P. B. Medawar

J. S. Medawar

HARPER COLOPHON BOOKS
HARPER & ROW, PUBLISHERS
NEW YORK, HAGERSTOWN, SAN FRANCISCO, LONDON

A hardcover edition of this book is published by Harper & Row, Publishers, Inc.

THE LIFE SCIENCE. Copyright © 1977 by Peter and Jean Medawar. All rights re-
served. Printed in the United States of America. No part of this book may be used
or reproduced in any manner without written permission except in the case of
brief quotations embodied in critical articles and reviews. For information address
Harper & Row, Publishers, Inc., 10 East 53d Street, New York, N.Y. 10022. Published simultaneously in Canada by Fitzhenry & Whiteside Limited, Toronto.

First HARPER COLOPHON edition published 1978

ISBN: 0-06-090609-X

78   79   80   81   82   10   9   8   7   6   5   4   3   2   1

to

KARL POPPER

Contents

Foreword

This is a book about *ideas*, with no more factual information than is necessary to make the ideas intelligible. It is therefore in no sense a textbook: for one thing there are no diagrams of the insides of animals, and for another some of its content is too advanced for real beginners. Nevertheless, we think the book is a useful companion for genuine students of biology. It may also interest sociologists, anthropologists, philosophers, psychologists and literary folk who want to learn something of the conceptual framework of modern biology.

Plants, we admit, get a rather raw deal: neither of us is a botanist, but although botany has made hugely important contributions to physics — Brownian movement and osmotic pressure are evidence enough of this — the fundamental ideas of modern biology are mainly the work of zoologists and microbiologists.

The philosophical colour of the book — as it appears, for example, in our attitude towards definitions and a general impatience with irrationalism — reveals everywhere the teaching of Sir Karl Popper F.R.S. to whom, as a token of gratitude, our book is dedicated.

The philosophic opinions of M. Jacques Monod are referred to several times in the text. These are cited from his *Chance and Necessity* (Collins, London, 1972).

We are deeply grateful also to the foundations and scholarly institutions who provided us with the environments that made it possible to write this book: in particular, the Aspen Institute for Humanistic Studies, Colorado, and the Rockefeller Foundation's elegant retreat on the hillside overlooking Lake Como — the Villa Serbelloni.

Some difficult or unfamiliar ideas are explained each time they occur. These repetitions are intentional and will help to diminish the turning of pages to and fro. The Glossary will help when a technical term is used that is not fully explained until a later chapter.

Everybody knows that the piecing together and preparation for press of a book such as this makes the most exacting demands upon the patience and skill of secretaries, and we feel specially grateful to Joy Heys and Valerie Price for the friendly and efficient way in which they have coped with the preparation of the text for press.

P.B.M.
J.S.M.

The Life Science

Chapter One
Introduction

Biology is a general term comprehending all the Sciences (the 'Life Sciences') that have to do with the structures, performances and interactions of living things. They include the conventional biological sciences of teaching syllabuses, viz. Botany, Zoology, Anatomy, Physiology and Genetics, and in addition the 'interfacial' sciences that have some of the characteristics of biological and physico-chemical sciences, viz. Biochemistry, Biophysics and Bioengineering, the last of which also establishes a common frontier between biology and communications theory.

Meanings of 'living' and 'dead'. Because it falls in with a common misunderstanding about the importance that should be attached to matters of definition, it would be quite understandable if everyone believed that biologists spent many hours debating anxiously among themselves the true meanings of the words 'living' and 'dead'. Such exercises would certainly take place if they served any useful purpose but, except in certain marginal cases mentioned below, they do not.

Discussions about (for example) the 'real' meanings of the words *living* and *dead* are felt to mark a low level in biological conversation. These words have no true inner meaning which careful study will eventually disclose. Laymen use the word 'dead' to mean 'formerly alive'; they speak only fancifully of stones as dead and never of crystals' living; but they can tell a living horse from a dead one and, what is more, can remember an apt metaphor that turns on the distinction. See N.W. Pirie, 'The meaninglessness of the terms life and living' (*Perspectives in Biochemistry*, 11, Cambridge, 1937).*

Being alive is a system property, i.e. a characteristic that can be attributed only to an organized system. This does not mean, however, that parts of such a system may not also enjoy the same property, for it is characteristic of living things that they are hierarchically organized; thus societies are made up of men and women, and individual human beings are made of organs which have a certain wholeness and functional unity of their own. Organs and tissues in their turn are composed of cells which also have a high measure of autonomy. It is perfectly understandable that a society should die — that is disintegrate — before any of its individual members and that a human being should die before one or other of his organs — e.g. the kidney — has lost its capacity to work in another human being as it did in its original owner. Transplant donation would be impossible but for this dispensation.

The marginal cases mentioned above never turn upon mere squabbles about the meanings of words but upon profoundly important matters of hard empirical fact. One such example is the anxious question of whether or when a potential transplant donor (e.g. a kidney donor) can be regarded as dead. The question at issue here is whether or not the changes which the potential donor has undergone in becoming insentient and unable to keep going are indeed irreversible. This decision is clearly a very important one; on the other hand, nothing turns on whether a virus is described as a living organism or not. Some properties of viruses — for example their extractability, infectiousness and mutability — tempt us to think of them as minute living organisms smaller even than bacteria, but other properties, especially their ability to subvert the synthetic machinery of a living cell in such a way that they will produce more copies, make us think of them as no more than neat packages of genetic information. So far from being 'alive' a virus is

---

* This paragraph is quoted from P.B. Medawar, *The Future of Man* (Methuen, London, 1960), p.108.

simply a piece of bad news wrapped up in protein.

Protoplasm. It was at one time quite widely believed that the material basis of all life was a complex substance to which the name 'protoplasm' was given. Protoplasm was thought to be a sort of fragile biological ether permeating otherwise inanimate structures, much as the physicist's hypothetical ether permeated all material objects. Both turned out to be equally mythical. When Thomas Henry Huxley described protoplasm as the physical basis of life, it was the word 'physical' that carried most weight. Life or liveliness was henceforward to be regarded as something that could be enjoyed by objects that were physical objects in a conventional sense: living things did not depend for their existence upon the operation of a life-force or vital spirit of any kind. Indeed, since Huxley's day the doctrine that J.H. Woodger has described as 'dogmatic vitalism' has steadily lost ground and is no longer seriously entertained by biologists. Dogmatic vitalism declares that living systems consist of their material parts in a conventional sense *plus* something else that endows them with vivacity.* It is as if a motor car, in addition to wheels, brakes, gears, engine, etc. in the right functional relationships, had a spirit of mobility which empowered it actually to run. There are many unsolved and deeply puzzling problems about how living things work and how they keep going, but no modern biologist feels that these problems are made in any way more easily soluble by having recourse to notions such as *entelechy*, or *élan vital.* The view antithetical to traditional vitalism is sometimes called 'mechanism' and this too has a dogmatic and a temperate form. The dogmatic form of mechanism positively declares there is nothing whatsoever about living things that is not fully explicable in terms of the properties of their constituent parts — a view we shall return to when we discuss 'Reducibility' and 'Emergence' (Chapter 22).

There is no such thing as 'protoplasm' and the word is used in professional writing either as a conscious archaism or as a

---

* 'Vivacity' is an attractive old-fashioned name for what is nowadays rather pallidly called 'vitality'.

figure of speech. Yet so strong a hold did the idea of proto-plasm have over the minds of Victorian biologists that Thomas Henry Huxley, in a notorious paper in the *Quarterly Journal of Microscopical Science*, gave a detailed circumstan-tial account of the properties of an organism dredged from below 2,000 fathoms in the North Atlantic and consisting, so he declared, of virtually naked protoplasm. The new organism was named *Bathybius haeckeli* as a mark of respect for Ernst Haeckel, the great zoologist of Jena, who first proposed the existence of a group of organisms, the *Monera*, consisting essentially of *Urschleim* (primitive slime). The entire episode makes a good example of 'poetism', i.e. the adoption of a scientific theory because of its elegance, attractiveness or romantic appeal. (See P.B. Medawar, *The Hope of Progress* [Wildwood House, London, 1974], pp.128–30.) Another example of poetism is the profoundly unsound notion that human blood, and the blood of land vertebrates generally, was in some sense an evolutionary descendant of a body fluid that represented an entrapment of the primeval oceans in the era when vertebrate animals left the sea to colonize dry land. Many nature-philosophers have found it profoundly appealing that, in addition to much other evidence of a remotely distant origin, human beings should carry round with them a memento of the Silurian sea.

For many years the mystique of protoplasm lingered on in the belief that life might be a manifestation of the behaviour of some complex, exquisitely well-balanced colloidal system. Indeed, Sir Frederick Gowland Hopkins, the father of bio-chemistry in Great Britain, said at a meeting of the British Association for the Advancement of Science in 1913 that life is 'the expression of a particular dynamic equilibrium in a polyphasic system'. This view gained countenance when, thanks to the work of J.D. Bernal and others, the existence of 'liquid crystals' came to be recognized – i.e. the existence of crystalline orderliness in solutions. Today, however, it is no longer believed that colloid chemistry is a special sort of chemistry – that colloids have properties other than those to be expected of solutions of very large molecules that often bear electric charges. Indeed, the 'basis of life' – if such a phrase has any meaning – is structural in an almost crudely anatomical sense: molecular transformations occur in a certain sequence and in a certain place because the agencies through which they are mediated (mainly enzymes, see p.84) enjoy a certain orderly structural arrangement. Electron microscopic examination of cells reveals solid structures which have definite shapes and look as if we could pick them up and handle them if only they were large enough. As theories of vital organization of the protoplasmic genre have quietly disappeared from view, it is to the expert in high resolution electron microscopy rather than to colloid chemists that we now look for an understanding of the way things are ordered in biological systems.

Teleology and teleonomy. Biologists are well known to shun the notion of teleology in explaining biological phenomena, but to the layman it seems obvious — and quite rightly obvious — that purposiveness is one of the distinguishing characteristics of living things. *Of course* birds build nests in order to house their young and, equally obviously, the enlargement of a second kidney when the first is removed comes about to allow one kidney to do the work formerly done by two.

What then is thought so objectionable about this form of words? For Aristotle, the Final Cause was 'that for the sake of which a thing exists'. It is this connotation of teleology that gripes the biologist, that is to say the use of its purpose or the nature of the goal actually fulfilled as a *causal explanation* of any biological performance. The attitude of biologists to teleology is like that of the pious towards a source of temptation which they are unsure of their ability to resist. This is the reason why biologists prefer to use the genteelism *teleonomy* with merely descriptive connotations to signify the goal-directed or 'as-if purposive' character of biological performances. Mr G. Sommerhoff introduced the term *directive correlation* to describe the relationship between separate biological performances that must exist between them, if they are to be directed towards the achievement of a goal.* In this book, however, we shall several times use the word 'teleology' quite unashamedly; we do not

---

* G. Sommerhoff, *Analytical Biology* (Clarendon Press, Oxford, 1950).

feel steeped in guilt by enquiring into the teleology of graft rejection, which may represent a tiresome byproduct of a body-wide monitoring system that exists in order to spy out and eradicate malignant variants of the body cells ('immunological surveillance', q.v.).*

Specificity. When biologists talk about their work, it is not long before the notion of *specificity* crops up. In a biological context the word refers to a unique one-to-one relationship between an agent and that which is acted upon or an agent and the effect it produces. The paradigm of all specific relationships in biology is that which holds between antigen and antibody (see Chapter 13, Immunology), or more generally between an immunity-provoking agent and the response it excites; other examples of specificity are the unique relationship between stimulus and response in a simple reflex action and between an enzyme and the substance it acts upon (its so-called 'substrate'). Specificity of antigen/antibody and enzyme/substrate interactions has a molecular basis, i.e. depends upon the relative molecular configurations of the interacting agents, but the specificity of a reflex action may of course have a simply anatomical basis, just as the specific relationship between a train's declared destination and the place it actually gets to depends upon the way the tracks are laid.

The frequency with which the notion of specificity crops up is not of course the only way in which the ordinary

---

* The philosophic dangers of too supine a dependence on the notion of teleology are illustrated by Sir Ernest Gowers's citation of a ten-year-old schoolgirl's essay on the cow:

> The cow is a mammal. It has six sides — right, left, an upper and below. At the back it has a tail on which hangs a brush. With this it sends the flies away so that they do not fall into the milk. The head is for the purpose of growing horns and so that the mouth can be somewhere. The horns are to butt with, and the mouth is to moo with. Under the cow hangs the milk. It is arranged for milking. When people milk the milk comes and there is never an end to the supply. How the cow does it I have not yet realized, but it makes more and more. The cow has a fine sense of smell; one can smell it far away. This is the reason for the fresh air in the country ... the cow does not eat much, but what it eats it eats twice so that it gets enough. When it is hungry it moos, and when it says nothing it is because its inside is full up with grass. (Sir Ernest Gowers, *The Complete Plain Words* [Penguin Books, Harmondsworth, 1973], p.69.)

professional discourse of biologists differs from that of physicists and chemists: another is the variety of time-dependent notions and the frequency with which they too crop up. 'A snowflake is the same today as when the first snows fell,' said D'Arcy Thompson, and this invariance, characteristic of the physical world, distinguishes it very strikingly from the biosphere, where all is change, where the life cycles of conception, development, senescence and death are superimposed upon the grand secular transformations of evolution. The notion of change in time pervades the whole of biology. For reasons of professional pride and because they fear that to believe otherwise would be to give free rein to secularism and free thinking, besides increasing the likelihood that they will be put out of a job, many biologists like to believe that there is an absolute categorical distinction between the biological and physical sciences, but their beliefs have not been gratified. The operational distinction between them, clear enough to have satisfied all the universities in the world, populated though many of them are by pedants, is that biologists study living organisms or their parts while physicists and chemists study inanimate systems.*

Information, order and noise. The idea of 'information' crops up repeatedly in the course of this book: a figure of speech that would not have gained such wide acceptance in biology unless it served a very useful purpose. 'Information' connotes order and orderliness or anything that embodies it — e.g. a message or set of instructions specifying order. There is nothing about this usage which differentiates it in principle from that which makes it natural to say that architects' drawings embody information for building one sort of house rather than another. 'Genetic information' is the kind of orderliness found in the structure of nucleic acid molecules which specifies or embodies instructions for the assembly of molecules in one particular way.

The connection between information and order is a perfectly straightforward one and reminds us of the fact that the notion of information was imported into biology from

---

* The late Professor C.F.A. Pantin, F.R.S. has written interestingly and convincingly on many of the topics discussed here. See C.F.A. Pantin, *The Relations between the Sciences* (Cambridge University Press, Cambridge, 1969).

communications engineering. Clearly the information in a message — e.g. a telegram or a letter — depends upon the order of the words and the order of the letters in the words: the more numerous the possibility of recombination among them, the greater the information *capacity* of the letter or telegram. 'Noise' connotes the very opposite of information, that is to say randomness or disorderliness. A signal which conveys no information in a context in which it might have been expected to have done so is dismissed as noise. Such a description applies both technically and in the everyday sense to, for example, the crackles and strange ethereal murmurings that sometimes accompany radio programmes, and it applies also to the snowstorms which capriciously interrupt TV programmes. When the noise signals are so subdued, random and heterogeneous that their pretensions to conveying information are negligible, we may speak of 'white noise', e.g. the sound — as of innumerable mice eating Rice Crispies — that sometimes accompanies long-distance telephone calls. All noise represents the intrusion of a disorderly element into an informational system.

Among the many ways in which it is possible to express the truth embodied in the second law of thermodynamics is to say that the overall direction of the flow of natural events in the universe is towards increasing randomness or disorder: an equipartition of heat throughout the universe would represent the ultimate dissolution of orderliness — the ultimate nonsense indeed.

It is often somewhat naively supposed that because the processes of development and of evolution entail an increase of visible order, living organisms must make use of some means of flouting or circumventing the second law of thermodynamics — a point discussed at length elsewhere.* Although living organisms do in reality obey it there is a sense in which they can be said to flout its spirit, for evolution and development are usually accompanied by an ostensible increase of orderliness and complexity. The point should not be pressed too far, however, because the increase

---

* See 'Evolution and Evolutionism' in P.B. Medawar, *The Art of the Soluble* (Pelican, London, 1969).

of visible complexity during development is at least in part a mapping of one form of order into another — a mapping of the genetic information contained in the nucleic acids of the chromosomes into the different form of order represented by specific proteins, enzymes and bodily structures (see Chapter 10, Bodily Constituents). In any event the second law applies only to *closed* systems — systems in which there is no external trade in matter or energy. All living organisms are thermodynamically *open* systems and their increase in order-liness is paid for by a net increase of disorder in their environments. (Refer here to E. Schrödinger's admirable *What is Life?* [Cambridge University Press, Cambridge, 1944].)

Cybernetics and feedback. Among the most fertile ideas introduced into biology in recent years are those of *cybernetics* — Norbert Wiener's general term for control theory or steering theory as it occurs in both engineering and biological contexts.

Control theory obtrudes everywhere into biology — in relation to body temperature, the salinity of the blood, blood pressure, pulse rate, the nice adjustment of hormone levels and so on.

'Feedback' — a term in constant use in biology — refers to one of the fundamental strategems of control. Feedback is the control of performance by the consequences of the act performed. Negative feedback, in its most familiar form, is exemplified by the switching off of a source of heat when temperature has reached a certain level, or in a biological system by a hormone's switching off the production by the pituitary gland of the trophic hormone (see Chapter 17, Co-ordination Systems) that would have stimulated its formation. At a characteristically noisy cocktail party, people raise their voices in order to make themselves heard, and this adds to the prevailing din so that people must shout louder and louder in order to make themselves heard at all. This continues until eventually they say to themselves, 'I can't *bear* cocktail parties,' and take their leave. This is an example of positive feedback, a fundamentally unstable and in extreme cases self-destructive process. In biology some forms of *autoimmune disease* (see Chapter 13, Immunology) illustrate the self-defeating character of positive feedback. When an immunologically self-destructive process is started up by injury to an organ, then the immunological process will itself cause more damage, which in turn acerbates the immunological response. Negative feedback enters into logic and scientific methodology in the process by which a hypothesis is rectified or modified by the nature of its logical 'output', i.e. by the degree of correspondence with reality of the theorems or predictions that follow from the hypothesis — a parallel which brings clearly to light the 'steering' element in scientific enquiry, which can be thought of as a means of finding one's way about the world. These examples are enough to show that the notions of cybernetics are pretty well ubiquitous in their application.

Cycles. It is not only in respect of hierarchical organization, time-dependence and the pervasiveness of the idea of specificity that the biological sciences differ so markedly from the physical. Another notion that pervades the whole of biology at all levels is that of *cycles.* The biological process in general is a story of cycles within cycles within cycles. The grandest cycles are of course the cosmic, i.e. the seasonal and diurnal, and the activities of living organisms fall in with them and adopt their pulse rate. Many biological cycles are restorative and regenerative, and the most important is the cycle of birth, maturation, reproduction, senescence and death. In micro-organisms such as bacteria and in many cells the reproductive cycles are continuous, but in larger organisms they often take their cue from the seasons.

Nothing illustrates more clearly the regenerative and renovatory characteristics of cyclical processes in biology than the grand cycles of use and reuse, synthesis and degradation, in which the elements nitrogen, carbon, oxygen and phosphorus — the elementary constituents of the biosphere — take part. Nitrogen forms about eighty per cent of the atmosphere and its compounds are essential ingredients of all living things. Nevertheless, there is no direct capture of atmospheric nitrogen by living organisms except by certain bacteria which live in symbiosis with leguminous plants. The commercial fixation of nitrogen in the manufacture of artificial fertilizers in quantities of the order of millions of tons per year is therefore extremely important.

Nitrogen compounds pass from one organism to another in the food chain and gaseous nitrogen is eventually returned to the atmosphere by the 'denitrification' that accompanies the final stages of breakdown and putrefaction of living organisms. In advanced industrial countries there is a serious leakage of nitrogen out of the cycle through wastage or misuse of sewage, which is specially rich in nitrogen compounds.

The *oxygen* and *carbon* cycles are closely interconnected. The traffic of both cycles passes through the atmosphere as the gases carbon dioxide and oxygen (they form 0.03 per cent and 20 per cent of the atmosphere respectively). Linked chains of carbon atoms form the backbone of every structural molecule in the body (see Chapter 10, Bodily Constituents) so perhaps it is not surprising to learn that more carbon is locked up in the form of coal and other fossil fuels than in living organisms themselves. This carbon is returned to the atmosphere by combustion, a process that uses up oxygen and converts the carbon into carbon dioxide. The linkage of carbon and oxygen cycles is the work of 'photosynthesis' — the process by which the energy of sunlight is used to transform water plus atmospheric carbon dioxide into the carbon compounds known as carbohydrates. This process is accompanied by the liberation of oxygen which is used directly in the respiration of both plants and animals. The most important — because they are the most abundant — organisms involved in the capture of carbon dioxide and the liberation of oxygen are forest trees and minute plants carried in the surface layers of the sea — 'phytoplankton'. At a certain depth in the sea the amount of carbon dioxide produced by respiration is just counterbalanced by the amount of carbon dioxide used up in photosynthesis — the equivalence zone.

Homology. One of the most important notions of biology is *homology*, the affinity of structure or behaviour that is to be attributed to a common origin, a parallel genetic determination or the relationship of being terms in an evolutionary series.

Starting from a simple five-rayed (pentadactyl) structure, the developmental processes that gave rise to the pectoral fins of fish have been transformed in birds into the developmental

processes that gave rise to wings and in mammals into forelegs. Pectoral fins, wings and forelegs are thus 'homologous'.
Another example of homology is that of the ear ossicles: the
tiny little bones that transmit vibrations from the eardrum to
the organ of hearing are homologous with bones that at one
time formed the suspensory system of the jaws. Likewise the
thyroid gland of mammals is homologous with an organ
which in lower chordates produced mucus and wafted it
round the inside of the branchial basket to trap fine food
particles in the current of water passing through it. In all
such examples it is important to remember that what has
taken place is not the evolution of one organ into another —
it is a mere naiveté to speak of the 'evolution of fins into
limbs' — but rather the evolution of one developmental
process into another. In the immediately post-Darwinian
era the elucidation of homologies and with it the consolida-
tion of the theory of evolution seemed to many biologists to
be their most important obligation, and it is still thought to
be so in those old-fashioned schools of zoology where the
teaching of comparative anatomy is the central core.

One of a beginner's very first lessons in comparative
anatomy is that the animals that were once rather loosely
grouped together as *Ungulates* — including horses, cattle,
sheep, pigs, giraffes and deer — are united by the property
of walking on the extreme tips of their toes, their hooves
being homologous with and of the same chemical composition as the nails (*ungulae*) or claws of mammals of other
kinds. The habit of walking on tip-toe has been accompanied
by modifications in the bony structure of the limb — 'hand'
or foot as the case may be. In ungulates, the joint ignora-
muses call the 'knee' is in the forefoot homologous with the
wrist and in the hind leg homologous with the ankle. Digits
of hand and foot have been reduced in ways that improve
their mechanical efficiency. The principal digit of the
horse's forefoot is that which corresponds to our own middle
finger, digits two and four being vestigial and no longer
reaching the ground, but in the forefeet of the even-toed
ungulates the remaining digits correspond to our own third
and fourth. It is easy to see how the investigation of homologies brings order into and makes sense of what might

otherwise seem to be a great untidy heap of zoological information.

The homologies of the limbs of ungulates are very elementary, and a more advanced zoological student — at all events in schools where comparative anatomy is rightly valued — will study such problems as the whereabouts of the front end of the vertebrate head — a good example of the kind of problem which to the layman seems comic, but which in reality turns upon investigations and reasoning of some formal elegance. The anatomically anterior end of the head is marked by the foremost of the segmental muscle blocks that occupy the vertebrate body from end to end, the foremost of the motor nerves that innervate them and the extreme anterior end of such longitudinal structures as the notochord and the nervous column — or alternatively organs which serve as landmarks of any of these, the neural element of the pituitary gland being one. Identification of the anatomical front end of the head makes it easier to understand the dramatic evolution of the higher nervous centres of the more advanced vertebrates in the form of a huge hypertrophy or overspill of neural elements *beyond* the anatomical front end of the head. The value of comparative anatomy is that it makes the system of animate nature easier to understand; it is a vulgarism to sneer at it and a self-inflicted punishment to disregard it.

Comparative anatomy no longer has the importance it once enjoyed, mainly because the greater part of the work has already been done — by the great German zoologists of the late nineteenth century, Gegenbaur and van Wijhe particularly, and their respectful and equally proficient British disciples such as Edwin Goodrich, who is said to have been the model for Dr Summerlee in Arthur Conan Doyle's *The Lost World* — in which Professor Challenger was unmistakably Professor Sir Edwin Ray Lankester, F.R.S. (1847–1929). The modern impatience with research as slow moving as comparative anatomy must not be allowed to distract attention from the fact that the study of comparative anatomy is an exacting and formally very beautiful discipline. Indeed, in the hands of some of its greatest practitioners it became almost a biological art form: a biologist who cannot appreciate and marvel at Edwin Goodrich's *Studies in the Structure and Development of Vertebrates* (Dover Press, 1930) deserves sympathy.

A note on microscopes. Although conventional microscopy opened up a whole new world of fine structure, the importance of 'ordinary light microscopy' and, in general, of *seeing* things instead of learning about them by other means must not be exaggerated. As explained later (p.29), the discovery of chromosomes and genes depended no more than the discovery of atoms and molecules upon our ability to *see* them; genes were known to exist long before the recognition of visible singularities in chromosomes that might be identified with them. Ordinary light microscopy has the disadvantage that nothing can be seen that is smaller than the wavelength of visible light. By using light of shorter wavelength — e.g. ultraviolet light and special lenses that are transparent to it — it is possible to see objects smaller still; the great revolution in microanatomy, however, came with the introduction of electron microscopy, which turns on the ability of highly penetrating electron beams to be focused and directed by magnetic fields very much as visible light can be lined up or made to converge or diverge by lenses. But electron microscopy has disadvantages of its own, chief among them being that material under examination must be bone dry and maintained *in vacuo* and that tissue slices for examination must be so thin that their preparation has become a work of very fine craftsmanship. Apart from these disadvantages, electron microscopy may be said to have disclosed a new world of crystalline orderliness within the living cell. For microanatomical purposes the most powerful electron microscopes,* with the highest resolving power, are not necessarily the best. Electron microscopes of only moderate resolving power are quite sufficient for much detailed microanatomy (rather than molecular anatomy), where the very highest magnifications would actually be unhelpful: in much the same way an observer can learn more

---

* The strength of a microscope is not measured by its power to magnify — for magnification can be merely 'empty', like a blowup of a coarse-grained photographic negative — but by its *resolving power*, i.e. its power to tell apart two objects or boundaries very close together.

about the shape and equipment of an approaching ship by using a pair of binoculars than by using a telescope powerful enough to show up the stubble on the captain's chin. In recent years it has proved quite easy to visualize the largest biological molecules, such as antibody molecules, by skilful electron microscopy and by using the highest resolving powers the structure of one virus — adenovirus 12 — has been almost completely elucidated.

Chapter Two
Biogenesis and Evolution

No principle of biology is more firmly established or less likely to be qualified than that of 'biogenesis', which avows that all living things are descended from living things. Behind each living organism today there is an unbroken lineage of descent going back to the beginnings of biological time. In its negative form the principle would state that there is no such thing as 'spontaneous generation' — e.g. the spontaneous generation of bacteria from putrefying organic matter or of protozoa from infusions of hay. Louis Pasteur, the greatest of all experimental biologists, is rightly credited with having carried out the experiments that falsified the notion of a spontaneous generation of bacteria and at the same time made an alternative hypothesis much more attractive, viz. that the bacteria which so readily proliferate in warm organic broths etc. derive from airborne organisms. This discovery, of which the medical significance was clearly perceived by Joseph Lister, lies at the root of all antiseptic and aseptic techniques in surgery today.

The principle of biogenesis applies not only to whole organisms but also to some of their constituent parts: among cellular organelles the *mitochondria* (see pp.127—8) are biogenetic in origin in the sense that they do not arise *de novo* by some synthetic process in the cell but are derived from pre-existing mitochondria only. Biogenesis does not imply evolution, but an evolutionary relationship does of course imply biogenesis. Normal biogenesis is often given the extra connotation of 'homogenesis', i.e. of like begetting like. Broadly speaking this particularization is true, although the theory of evolution obliges us to qualify it in detail. Thus the offspring of mice are mice and of men are men. No genuinely extravagant heterogenesis ever occurs, although in the days before empirical truthfulness was thought to be either a necessary or a desirable characteristic of professedly factual statements, all kinds of strange notions were rife — the most famous being the myth that geese might be born of such organisms as the attractive barnacle-like crustacean the goose barnacle, *Lepas anatifera*. Such notions belong to 'poetism', a style of thinking which arouses as much indignation among scientists as the more idiotic extravagances of computerized literary criticism arouse in lovers of literature.

Evolution and biosystematics. Samuel Taylor Coleridge once declared that zoology was in danger of falling asunder — the consequence of its huge mass of uncoordinated factual information. The evolutionary hypothesis* is that which brings an order and connectedness to what Coleridge saw as the great toppling heap of information that made up the zoology of his day. It can be regarded as an amendment to the biogenetic principle that like begets like (see above). The hypothesis states that the existing diversity of life-forms has arisen by progressive diversification during the course of biogenesis. It remains generally true to say that the offspring of mice are mice and of men are men, yet variants arise from time to time that may be recognized retrospectively as the beginnings of new specific forms. It is to the origin of these variants and the processes which keep them in being that we owe all the existing forms of life at present on the earth. Pedagogic 'proofs' of the past occurrence of evolution are of the same kind and unfortunately the same intellectual stature as those 'proofs' of the roundness of the earth which we learnt in our earliest schooldays. It is not upon these so-called 'proofs', however, that the acceptance of such a hypothesis depends. It is rather that the hypothesis of evolution pervades, underlies and makes sense of the whole of biological science in much the same way as the idea of the

---

* The word 'hypothesis' in this context is used in its correct and technical sense: it is a vulgarism to suppose that the word has a pejorative flavour and that in describing as a hypothesis what is usually called the 'theory' of evolution we are in some way depreciating it.

roundness of the earth permeates the whole of geodesy, chronology, navigation and cosmology. The evolutionary hypothesis is part of the very fabric of the way we think in biology. Only the hypothesis of evolution makes sense of the obvious inter-relationships between organisms, the phenomena of heredity and the patterns of development. For a biologist the alternative to thinking in evolutionary terms is not to think at all. Mechanisms of evolution are dealt with later.

The purpose of *biosystematics* is to name animals and to arrange their names in some order and pattern that will feel right even to biologists with rather coarse *taxonomic* sensibilities. Living things are classified in the first instance into 'kingdoms' — plant and animal — and, less monarchically, into 'phyla'.* The members of a phylum are united by a similarity of ground plan without regard to detailed differences of structure. A good example of a phylum is the *Arthropoda*, which includes crustaceans and insects, which resemble each other by having segmented bodies, an 'external skeleton' and multiply jointed limbs. Another fundamental similarity of structure is a nervous system that runs down the mid-ventral line of the body, with a ganglion in each segment that gives off branches towards the limbs. In addition, the blood vascular system is of the kind described as 'open' because the blood, which has only a minor respiratory function, does not run in anatomically well-defined channels such as arteries or veins but rather percolates through the tissues of the body until it returns to a heart which occupies a dorsal position, in contrast to its ventral position in vertebrate animals.

Invertebrates** tend to group themselves, on the one hand, into phyla such as the Arthropoda, the worms rightly so-called (*Annelida*, including the earthworm, whose busy beneficence and modest unobtrusiveness are an example to us all) and the worms wrongly so-called, that is to say the roundworms, eelworms (*Nematodes*) and flatworms, many of

---

* The broadest distinction of all is between organisms of which the genome is (*eukaryotes*) or is not (*prokaryotes*) organized in the form of compact chromosomes.

** 'Invertebrate' is a description, not a taxonomic term.

which are parasitic. On the other hand, in total contrast to arthropods, several invertebrate groups are related to the chordates,* including the vertebrates and therefore ourselves, by certain characteristics of very early development and by the possession of a roomy and often tripartite body cavity, the so-called 'coelom' which lies between the connective tissue of the outer body wall and the connective tissue which surrounds and supports the viscera. Groups belonging to this chordate line of descent — amongst which, however unlikely it may appear, we must expect to find the modern representatives of our own remotest ancestors — are echinoderms, including sea urchins, starfish and sea cucumbers, phoronids, sea-arrows and the large group known as sea-squirts, whose chordate affinities are so obvious to professional zoologists that they are classified as chordates anyway. Thus in the invertebrates generally one can identify two main streams of evolution and two great classes of affinity: that associated with annelids and arthropods on the one hand and with chordates and vertebrates on the other.

The taxonomic stature of phyla and classes has already been mentioned. Next in order after classes of animals come orders, families, genera, species and individuals; each major grouping can of course be further sub-divided into a group of subordinate status, e.g. a sub-phylum or sub-class or sub-species, but these need not concern us.

Fish, amphibians, reptiles, birds and mammals are examples of *classes* among vertebrates and insects, crustaceans and arachnids (spiders) are examples among arthropods.

Orders are meant to be of the same taxonomic stature or weight throughout the animal kingdom. Members of the same order obviously have a closer affinity to each other than that which is implied by their common membership of a class. Unfortunately, it is not possible to define this degree of affinity in a way that could be valid throughout the entire animal kingdom. Like many other decisions that depend mainly upon individual judgment, a taxonomic allocation is rather a matter of experience and 'feel' than anything that

* Chordates have characteristically a *dorsal* nervous system underlain by a simple undivided elastic skeletal rod: the *notochord*. The heart is ventral.

can be arrived at by consulting a rule book. Taxonomic disputes are usually conflicts of personal judgment and therefore often virulent and unforgiving. The stature of the category known as orders can be appreciated by reflecting that among birds, ducks and geese and swans form one order, *anseriformes,* turkeys and chickens a second, *galliformes,* and owls, *strigiformes,* a third; in the class of mammals, whales and dolphins form one order; beavers, chipmunks, squirrels, rats and mice a second; monkeys, apes, human beings, chimps and gorillas — all primates — a third. Within orders we recognize as *genera* animals that are very obviously 'of a kind', as the great cats are clearly of a kind. Lions, tigers, leopards, jaguars and panthers make up a genus. Using the nomenclature of the Zoological Society of London, *Panthera leo* is lion, *Panthera tigris* tiger, *Panthera pardus* leopard and *Panthera panthera* the panther commonly so called. These examples also illustrate the binomial nomenclature introduced by Linnaeus: the species is designated by a generic name — in these examples *Panthera* — followed by a second term which serves to differentiate, e.g. lion from tiger. It is a fully established convention of biology that species are referred to by both a generic and a specific term. The specific term is never used alone in a biological context. Tigers make up the species *Panthera tigris* and not the species *tigris.* Anyone who refers to 'the *amoeba*' forfeits all hope of being mistaken for a professional zoologist.

A species is a community of actually or potentially interbreeding organisms, which either for genetic or for geographic or behavioural reasons have achieved a sufficient degree of reproductive isolation to enjoy the possession of a distinctive make-up and frequency of genes. This population-genetical definition of a species has many practical drawbacks in spite of its theoretical attractions. If the characteristic of a species is the possession of gene X by fifty-five per cent of its members and of gene Y by ninety per cent, the idea of an *individual*'s belonging to a species becomes a bit vague except in a probabilistic sense, for only populations can really be or not be members of species.

A hard-working museum taxonomist can be driven out of his mind by the assurance that the problem of species definition has now been solved: a species is essentially a cluster of points in *n*-dimensional character-space.

Insects.   Although insects are the great success story of evolution, Darwin complained that entomologists, the people who study them, were the very last to be won over to his conception of the evolutionary process.

Insects are probably the most numerous and certainly the most various of all many-celled animals.   The only possible contestant on grounds of number must be the minute crustacean *Calanus* — a principal constituent of the *Plankton* of the surface waters of the oceans throughout the world.   Insects owe their success as a group to their high reproductive rate and genetic variance, which have made it possible for them to exploit almost every environment capable of supporting life except the sea, in which other animals of the same ground plan (i.e. arthropods), mainly the crustaceans, enjoy insect-like diversity and numerical preponderance. Perhaps, then, it is the arthropodan ground plan that has been so successful.

Insects, it has already been explained (p.24), have an 'open' blood vascular system, and like crustacea have the hard outer casing — an 'exoskeleton' — that goes naturally with it; without this hard outer casing the contraction of the heart would simply cause a bulge on the surface instead of a propulsion of fluid.

The fact that an open blood system creates little in the way of turgor pressure accounts for the characteristic floppiness and shapelessness of the internal organs of insects and crustaceans — which contrasts so sharply with the firm roundness of the internal organs of vertebrates.

The possession of an exoskeleton has other important implications for insects and arthropods generally.   In vertebrates, closed bony boxes like the cranium can grow only by the deposition of bone on the outside accompanied by its removal from the inside.   In arthropods no such process is possible, so bodily growth must be accompanied by periodic shedding — moulting — of their hard outer casing.

Many insects go through larval forms that occupy an important fraction of their total life cycle.   These larvae, e.g. the caterpillars of butterflies and moths, undergo a profound internal reorganization (pupation) during their reshaping into the adult.

In some insects the larval stage is the greater part of the life cycle, and the ephemeral adult into which they transform, e.g. the mayfly, which does not feed, is hardly more than an airborne reproductive organ.

It is generally taken that insects evolved from organisms akin to annelids (p.24), and that the present-day insects most faithfully representative of the evolutionary prototype are those possessing the most 'generalized' structure: the *orthoptera*, which include cockroaches and grasshoppers, answer this description well. In the heyday of evolutionary biology there was thought to be an obligation upon entomologists to trace out, as far as possible, all the lines of evolution within insects, but most modern entomologists have given up this activity as tiresome and fruitless: nothing of any importance turns on the allocation of one ancestry rather than another.

The study of insects abounds with interesting and important biological problems to do with heredity, development, behaviour and the action of hormones. The gaseous exchange in their respiration, for instance, is mediated through very fine air tubes, *tracheae*, leading directly from the atmosphere to the internal organs. The inherent physical limitations of this respiratory system, combined with the necessity for moulting, set a limit to the size of insects which is much more exigent than that which applies to, for example, crustaceans. Thus a repopulation of the world by huge, fascist insects may be regarded as one of the more idiotic Gothic extravagances of science fiction; moreover, the likelihood that insects will evolve into animals of any other kind may be dismissed as negligibly small. Insects are specialized endproducts of evolution. It has been said that within the group itself new species may be originating faster than they can be recognized and named. Such a claim can hardly be verified, however, because modern entomologists are no longer engaged in merely taxonomic exercises. What *is* quite certain, however, is that not all living and fossil species have yet been described and named, although about a million are known. We can also be sure that the adaptive finesse responsible for the success of insects has closed the door to new evolutionary possibilities.

Chapter Three

Biological Inheritance,
Nucleic Acids, The Genetic Code

In everyday life we do (or more often do not) inherit wealth, property, honorific titles, or personal possessions of various kinds. In biological inheritance, however, what we inherit is a chemically coded message — a coded set of instructions — which specifies very exactly the course which the development of the next generation of organisms is to pursue. The chemical molecules that encode this genetic information are the giant polymeric molecules of deoxyribonucleic acid (DNA for short). The information-carrying capability of nucleic acid molecules depends on the virtually inexhaustible combinational and permutational variety of the four different nucleotides out of which the molecule is compounded (see Chapter 10, Bodily Constituents and Chapter 12, Molecular Biology). The information transfer system may thus be likened to a morse code with four different symbols instead of the familiar two-symbol dot and dash. The nucleic acids are transported within, and form a major part of, the substance of the chromosomes in which the nucleic acids are stabilized by combination with a basic protein. Chromosomes are the material objects transmitted in the sperm and egg cells from one generation to the next; they are differentiated along their length and the singularities of nucleotide order which distinguish one stretch of chromosome from any other are known as 'genes'. Chromosomes are visible under an ordinary microscope and the singularities which represent genes are now beginning to be identified by electron microscopes of sufficiently high resolution. It is noteworthy, however, that even if the microscope had never been invented and the biologist had no means of magnification at his disposal, we should still know of the existence of chromosomes and genes by reasoning of the same kind as that which has allowed us to believe in the reality of molecules and atoms, but in this case the reasoning is purely genetic, i.e. has to do with the outcome of breeding experiments. It is, indeed, still by this means that the existence of most genes is known to us. Some genetical purists of an older tradition, anxious to establish the autonomy of genetics and to protect themselves from the pretensions of molecular biology, are fond of emphasizing the degree to which they would have been able to work out the material basis of heredity without recourse to physical or chemical methods of analysis. Geneticists of the generation of William Bateson (1861—1926), who did so much to promulgate and at the same time add conviction to Mendel's laws of heredity, were indeed sometimes even impatient of all the talk then current about chromosomes. The elementary account of Mendelism which follows is still essentially Bateson's, although it is reworded to bring in chromosomes.

In the fertilized egg chromosomes are present in pairs — the *diploid* condition — one chromosome from each pair having been contributed by each parent. Inasmuch as the ordinary 'somatic' cells of the body are derived from this fertilized egg by a succession of symmetrical cell divisions, replicas of these chromosomes are present in pairs in virtually all the cells of the body. However, in the process of gamete formation (the gametes, 'germ cells', are sperm and egg cells), the chromosomes of a pair separate in such a way that only one chromosome from each pair enters each gamete; the number of chromosomes in each is therefore half that of the ordinary somatic cell — the *haploid* condition. When fertilization occurs in the union of gametes from individuals of different sexes, the diploid number is restored and the chromosomes pair up with their corresponding opposite numbers from the other sex. It should be emphasized that after parting the chromosomes are dealt out into gametes entirely at random: there is no knowing into which gamete any one chromosome of a pair will go and the chance of any one chromosome's going into one specified gamete is exactly the same as the chance of its opposite number's entering that gamete. Fertilization is also an entirely random process from a genetical point of view, but the mechanics of the process of chromosome segregation and reunion ensure that there is a certain statistical regularity in the apportionment of heritable characteristics among the offspring. These regularities are embodied in Mendel's famous segregation ratios (3:1; 9:3:3:1) of which an account may be found in any elementary genetics or biology text-book. It was not until after the turn of the century that biologists came to realize that the observable behaviour of chromosomes conformed pretty exactly with that which would be expected of any material agent responsible for propagating Mendel's 'hereditary factors'. Chromosomes corresponded to linkage groups, i.e. to groups of genetic determinants inherited together if at all, and the material basis of the genetical factors themselves came to be called 'genes' after the outstandingly brilliant work of Thomas Hunt Morgan and his genetical school at Columbia University had made it certain that chromosomes are differentiated along their lengths. Thanks to Morgan's school a new and highly important source of genetical variegation was uncovered — one that adds very greatly to the combinational varieties of the genes that may be found in the different individuals of a species.* This is *crossing over*, a process in which genetical material is exchanged between the chromosomes of a pair, so breaking down conventional linkage groups. Thus just as the phenomenon of linkage may be thought of as a qualification of the principle of the independent assortment of genes into gametes, so crossing over may be thought of as a qualification of the phenomenon of linkage. Crossing over, independent assortment and the random recombination of gametes provide between them for the enormously broad spectrum of genetic diversity which distinguishes every outbreeding species of organism.

It is a fundamental characteristic of Mendelian inheritance that genetic determinants maintain their integrity generation

---

* The combinational varieties of known human genes outnumber all the people who are alive today, have ever lived or are likely ever to live.

by generation, so that genetic determinants temporarily lost in a cross may be recovered many generations later in their original form and exercising the same effects. This is a genetical expression of the quite extraordinary molecular stability of the nucleic acids, i.e. of their comparative immunity to the disruptive influences which might be expected to affect molecules of this great size. Schrödinger,* the great theoretical physicist, was the first to call attention to the important consequences of chromosomal stability and was the first, too, to describe chromosomes as the vehicles of a genetic code, although the description of the gene as a message as opposed to an enzyme or model or agent of some other such kind was first used by Hans Kalmus.**

The English mathematical geneticist R.A. Fisher argued that Mendel had worked out virtually the whole of his theory on aprioristic grounds before he turned his mind to breeding experiments, so implying that his experiments were essentially of the kind that has been classified as 'Aristotelian',† i.e. were experiments intended to illustrate a principle already known rather than to uncover qualitatively new information or to test a hypothesis. Fisher's own intention may have been to illustrate the folly of supposing that Mendel's laws were 'deduced' from empirical observations on·plant breeding rather than devised by the 'hypothetico-deductive method', in which a preconception of what the truth *might* be precedes an experimental attempt to find out whether or not the preconception corresponds with real life.

There is certainly some substance in Fisher's argument because numerically speaking the results of Mendel's breeding experiments were a little too good to be true, i.e. came closer to a theoretical 3:1 or 9:3:3:1 ratio of character-distributions than is at all likely on the grounds of probability theory. It is possible that, with the kindest possible motives, Mendel's gardeners and other assistants gave the reverend scientist the results they suspected he would like to hear.

---

*   E. Schrödinger, *What is Life?* (Cambridge University Press, Cambridge, 1944).
** 'A Cybernetical Aspect of Genetics', *Journal of Heredity*, 41, 19 (1950).
†   See P.B. Medawar, *Induction and Intuition in Scientific Thought* (Jayne Lectures, American Philosophical Society, Philadelphia; Methuen, London, 1969), pp.34–8. ¹

One of Mendel's most brilliant achievements was to recognize and comprehend within his theory the phenomenon of *dominance*. A gene is said to be dominant when it can express itself in spite of being inherited from only one parent, i.e. when present in the heterozygous (see below) state; conversely, genes that can express themselves only in the homozygous state, i.e. when inherited from both parents, are said to be *recessive*. Dominance and recessiveness are properties, not of genes, but of the way in which they express themselves, which may be influenced by many other genes in the genome. In the simplest examples of dominant and recessive gene pairs — e.g. genes affecting our ability or inability to taste the chemical compound phenylthiourea — the dominant gene completely masks the effect of the recessive, for the ability to taste phenylthiourea leaves little room for the expression of an inability to do so. Blood group genes, however, are said to be 'codominant'. Codominant genes express themselves even when paired with genes of a different kind. Such is the case of members of the blood group AB. If our methods of analysis were sufficiently sensitive, recessive genes could be identified even in those heterozygotes in whom their effects were masked by the action of dominant alleles. It is indeed a matter of considerable importance in eugenics (q.v.) that the carriers of harmful recessive genes *can* sometimes be identified.

In the simplest early formulations of Mendelian heredity, it was supposed that all hereditary determinants existed in binary alternative forms — alleles. Their alternative manifestations were sometimes as simple as the mere presence or absence of a particular character trait, such as (in human beings) the ability or inability to taste the chemical compound phenylthiourea, or the alternative forms made famous by some of Mendel's early experiments, e.g. green as opposed to yellow peas or round as opposed to wrinkled peas. When genetic determinants of the same kind are inherited from both parents a subject is said to be *homozygous* with respect to that determinant. People who are homozygous for some character trait produce gametes which all have the same determinant in respect of it. When dissimilar genetic determinants are inherited from the two parents the

subject is said to be *heterozygous* in respect of that characteristic. Heterozygous individuals produce gametes of different
kinds, because the chromosomes which are the vehicles of the
corresponding unlike determinants separate and enter different gametes. The members of ordinary, outbreeding, natural
populations are unlike, hybrid or heterozygous for most
character traits. It is indeed very unlikely that *complete*
homozygosity for all genetic determinants can be achieved in
any naturally outbreeding population, although an approximation to it can be arrived at by systematic and prolonged
inbreeding (see below).

Inbreeding and hybrid vigour.   The genetic systems of
most animals — and in human beings their customs and civil
laws as well — provide for *outbreeding*. Inbreeding is a
systematic pairing, generation by generation, of genetic
relatives and in the most extreme form, where it is possible,*
is the self-fertilization of hermaphrodite organisms.

The effect of prolonged inbreeding is to change the genetic
make-up of a population in the direction of homozygosity
and ultimately of genetic uniformity: the mice used in many
cancer research and immunological laboratories have been
developed by a programme of inbreeding that has involved
the mating of brothers and sisters or parents and offspring for
upwards of twenty successive generations. Such mice are
homozygous for most gene pairs and one of the ways of
confirming their genetic homogeneity is to show that they
will accept skin transplants one from another. In the course
of becoming homozygous deleterious recessive genes will
often be brought into conjunction, sometimes with fatal
effects. For this reason most attempts to start inbred lines of
mammals, which are naturally outbreeding, end in failure. In
human beings it is well known that rare disorders of recessive
determination turn up much more frequently in the offspring
of consanguineous matings such as cousin marriages: the rarer
the disorder the greater will be this relative preponderance of
hereditary disorders among the offspring of relatives. Communities that practise some measure of inbreeding may be
assumed to have excreted highly deleterious genes present in

* Hermaphrodite animals such as snails and earthworms invariably cross-fertilize.

the gene pool of the community as a whole — i.e. the unfortunate people in whom their effects were expressed will have
died.

When two inbred strains of mice are crossed their hybrid
progeny are livelier, healthier, more rapidly growing and
larger mice which are normally very much more resistant to
disease and all other forms of stress than the inbred lines
from which their parents were derived. This hybrid vigour,
'heterosis', is a phenomenon very familiar to all stock-
breeders. Its effects could be due to the masking of the
action of rather unfavourable recessive genes by their normal
opposite numbers or counterparts — unless, of course, the
same harmful recessive gene should happen to be fixed in
both lines. A more general consideration is that natural
populations of outbreeding animals are highly heterozygous
in make-up, so that the pattern of action of each gene will
have evolved in relation to an essentially heterozygous
constitution such as the hybrid provides.

The experience of stockbreeding has created the popular
fallacy that human beings who are the product of inter-racial
crosses are especially well endowed in respect of physique,
beauty, intelligence and sexual prowess — though not of
course in *gentility*, for it is part of a cognate system of myths
that true gentility, and with it the moral virtues of honesty,
courage and fortitude, can only come from 'pure breeding'.
There is no real substance to any of these beliefs because all
natural populations, including human populations, are
heterozygous and heterogeneous and the genetic make-up of
each one will be that which has adapted them most effectively to the environments in which they live. There is no reason
therefore why the progeny of a cross between members of
two such populations should be specially well endowed.

Nature and nurture: genetics and epigenetics. Except
avowedly as a means of avoiding all kinds of long-winded
periphrases, expressions like 'the gene for albinism' or 'the
gene for tallness' are nowadays judged to be clumsy and
potentially misleading, for their wording takes no cognizance
of the *epigenetic* element in development. Genetics deals
with the character of the information transmitted from
generation to generation, 'epigenetics' (Waddington's term)

deals with the processes by which this information is trans-lated into real life — into flesh and blood and distinctive behavioural characteristics. No modern biologist would ever speak of 'the inheritance of intelligence', or of a genetic make-up endowing its possessor with intelligence, because with intelligence the importance of the epigenetic, nurtural element of upbringing is so very widely known and appreci-ated. It is however safe to speak of inherited *differences* of intelligence or differences of stature because such a formula-tion envisages the environment or the developmental ambi-ence as a parameter in the developmental equation while the genetic contribution varies. The same formulation may also be recommended when dealing with other inborn differences. For example, rather than speaking of a gene 'for' blood group A or B as the case may be, it is better in general to speak of genes that control blood group *differences*. However, it matters less in this context than any other because the genes that control blood group differences exercise their character-istic effects in almost any environment that is capable of supporting life.

It was at one time naively believed that in respect of each character trait it was possible to specify a precise degree to which its manifestation was due to 'nature' or to 'nurture' — to inheritance or to environment — thus it would be gravely asserted that intelligence was due seventy-five per cent to heredity and twenty-five per cent to environment. Such a formulation is utterly unacceptable: as Hogben and Haldane took pains to make clear, the degree to which a character difference may be attributed to the effects of nature is itself normally a function of nurture and *vice versa*. Thus our human susceptibility to scurvy might have been regarded as an 'inborn' character trait in an older way of looking at things, but when we reflect that such a supposedly hereditary trait can make itself manifest only in an environment lacking in vitamin C we conclude that a categorical distinction between the operations of nature and of nurture cannot really be drawn. The fact that this is so should not, how-ever, be allowed to derogate from the importance of attempting to define the functional relationship between nature and nurture in, for example, disabilities such as diabetes.*

Twins. It is in such a context as this that the study of twins is specially important. Human twins are of two kinds: fraternal and identical. (a) Fraternal twins are litter mates — human beings do sometimes have litters of two or more. In a genetic sense they resemble each other no more closely than ordinary brothers and sisters and they are not necessarily of the same sex. Fraternal twin pairs are also called 'dizygotic twins' because each of them is formed from a separate fertilized egg or zygote, whereas (b) identical twins are formed from a single zygote and begin life as a single individual. Becoming two or more is the consequence of a division of the zygote into two daughter cells which separate, each half becoming a separate embryo — a division that may occur at a somewhat later stage, though needless to say this power to readjust is lost very rapidly in the course of development. Barring chromosomal accidents, offspring of monozygotic origin are genetic replicas each of the other. The judicious use of identical and non-identical twins therefore makes it possible to distinguish with particular clarity between differences due to nature and differences due to nurture: when identical twins are reared apart — something that happens more often than is generally supposed, though less often than medical geneticists would like — they will have been exposed to entirely different sets of 'nurtural' stimuli, and the differences between them may thus fairly confidently be attributed to differences of upbringing in a wide sense. Conversely, when dizygotic twins are brought up together but yet diverge in characteristics, the differences between them may reasonably be thought to be genetic in origin. In practice, the formally exact situation in which identical twins are reared apart from a very early age or dizygotic twins are reared under exactly similar conditions can very seldom be realized on a large scale, so that medical biologists have recourse to the method of concordance or discordance. Thus if our intention should be to discover to

* The use of 'nature' and 'nurture' to distinguish inborn differences from differences due to upbringing is at least as old as Shakespeare: in *The Tempest*, Act IV, Scene 1 Prospero describes Caliban with quite undue asperity as 'a devil upon whose nature nurture will never work'.

what degree susceptibility to diabetes is due to inherited differences or to differences of upbringing, it will be helpful to examine the relative frequency with which diabetes occurs in both members of the twin pair when the twins are identical on the one hand or dizygotic on the other. If concordance is more common among identical twin pairs then we shall have reason to suspect the intervention of an important genetic element. If, on the other hand, concordance is just as common among dizygotic twins, we should be tempted to attribute susceptibility to diabetes to a nurtural rather than to a natural difference.

Geneticism is a word that has been coined to describe the enthusiastic misapplication of not fully understood genetic principles in situations to which they do not apply. I.Q. psychologists are among its most advanced practitioners, and it must be reported that some of their evidence on the relative contributions of nature and nurture to differences of intelligence — particularly in twins — has come under suspicion of having been 'fiddled'.

Genetic systems. The 'genetic system' of a species is the general name for the sum total of the dispositions that provide for and affect the transfer of genetic information from one generation to the next. In describing the genetic system of a species we shall therefore specify the form of sexuality, the mating system — whether of outbreeding or of some degree of inbreeding — the mutation rate, the prevalence of crossing over — and any other factor which, varying perhaps from species to species, will affect the character of genetic information flow.

We are indebted to C.D. Darlington for the clear recognition of the fact that the genetic system of organisms can itself be a subject of evolutionary change. The genetic system of a species, whatever it may be, is therefore an endowment of adaptative origin. It is easy to imagine, for example, the disadvantage of hermaphrodite organisms in which self-fertilization was possible in comparison with organisms of the same kind which had evolved into possessors of devices that enforce outbreeding.

Sexuality

Sexuality is the state of affairs in which reproduction is a co-operative enterprise between two individuals; each contributes a germ cell or gamete, which fuse to form the fertilized egg or zygote from which development begins. The sexual process enormously enlarges the genetic variance of a population for each individual's genetic make-up is only one possible realization of a vast number (of the order of 10^{3000}) of possible combinations among the genes present in an interbreeding assembly of organisms. Thus an individual's candidature for natural selection is enormously enhanced. The arrangement that normally provides for an intermixture and commingling of genes from different individuals is usually a genetically enforced separation of the population into two sexes, each bearing complementary reproductive organs so that cross-fertilization is inevitable. When, as often happens in sedentary organisms, both egg cells and sperm are manufactured by a single individual, an 'hermaphrodite', then there is normally some special provision to make sure of cross-fertilization — some mechanism which ensures that an egg cell can be fertilized only by sperm from an individual other than that which produced the egg. Self-fertilization would clearly be a genetically self-defeating process and have all the disadvantages of inbreeding with its steady progression towards the extinction of genetic diversity. Full-blown sexual dimorphism is the simplest example of a very important kind of genetic polymorphism: in this case sub-division of the population into different genetic types that are mutually dependent upon each other and make no functional sense in isolation.

Sexuality in some form exists at almost all evolutionary levels, even in bacteria. In some bacteria, however, the sexual process is restricted to something that amounts to hardly more than the infection of one bacterium by nucleic acid from another — the 'parasexual' process that lies at the root of the enormously important phenomenon of bacterial transformation (see Chapter 11, Microbiology).

The *teleology of sex* has been made clear by implication in the paragraphs above: its purpose is to provide for genetic intermixture and so to enlarge enormously the repertoire of genetic make-ups which may be realized in a single individual.

Sexual cycles. Most mammals have an oestrous cycle in which the female alternates periods of receptivity with periods in which copulation is not allowed to occur. Primates are distinguished by the possession of a *menstrual* cycle with no well-defined period of female receptivity. Gavin de Beer has ingeniously suggested that the receptivity of the female at all times, taken in conjunction with the well-known sexual exigence of males, is one of the factors that has made possible the monogamous relationship found in some primates.

Sex determination. Sexual differentiation leading to approximately equal numbers of the two sexes is enforced in mammals by a switch mechanism using two specialized kinds of chromosomes, the X chromosome and the Y chromosome. These are extra to the ordinary chromosomes — *autosomes*.

Individuals of the make-up XY are male and XX female. Males produce sperm of two different kinds in roughly equal numbers — X-bearing and Y-bearing — whereas accidents apart, all eggs produced by females carry an X chromosome. If a Y-bearing chromosome fertilizes the egg a male will develop; and if an X-bearing chromosome, then a female. Thus in mammals the male is the sex-determining or 'heterogametic' sex, whereas in birds it is the other way about. No special privileges are associated with being the sex-determining sex. On the contrary: one of its consequences is that harmful recessive genes that would probably be masked in females by a dominant allele may come to outward expression. The gene associated with haemophilia is one such example. In males all bodily cells have the sex chromosome make-up XY, in females XX. The fact that female cells thus have a double dose of whatever genes are carried on the X chromosome would raise some tricky problems in physiological genetics were it not for the fact that one X chromosome of each pair is silent, or in effect switched off, in half the cells of the female — the process known as the Lyons effect after its discoverer Dr Mary Lyons. Sex determination in the sense of ascertainment of sex is made possible by the fact that in some body cells the extra X chromosome carried by females is microscopically visible as a 'Barr body' under the right conditions. This distinguishing mark is also an eponym of its discoverer — Dr Murray Barr. Ascertainment of the sex of a fetus is made possible by sampling the amniotic fluid, but in spite of its enormous importance for livestock farming and the wealth of ingenious research devoted to the problem no method has yet been devised for separating male-determining from female-determining sperm so as to increase significantly the likelihood of getting one sex rather than another.

Sexual development. The chromosomal switch mechanism described above provides for the development and maturation of the primary sexual characters, the respective sex organs; but the secondary sexual characters, which include the formation of mammary glands in female mammals and a number of behavioural traits, are called into being by the action of sex hormones manufactured by tissue associated with the primary sex organs. Some degree of sex reversal can be brought about by flooding an organism with sex hormones belonging properly to the other sex. This sex reversal may also transform sexual behaviour, so each sex must have the potentiality for behaving in the way characteristic of the other.

So much of structure and function is adapted to ensuring the success of sexual reproduction that any attempt to modify it, for moral or prudential reasons, confronts an ancient and formidable natural order which must nevertheless be reshaped if human beings are to survive (see Chapter 8, Demography).

Chapter Four
The Genetical Theory of Evolution By Natural Selection

Darwin's theory of evolution by natural selection was propounded long before he or anyone else had any clear idea about the true mechanisms of heredity. By the early twentieth century, the Darwinian theory of evolution had acquired such an all-encompassing explanatory glibness that natural historians such as D'Arcy Thompson had become quite uneasy about it. Nowadays we should probably attribute this uneasiness to the realization that a theory which explains everything explains nothing, and indeed today we can see a general parallel between the explanatory facility of the older form of Darwinism, the doctrines of psychoanalysis and the Marxian interpretation of history. Nevertheless, Darwinism is the theory of evolution that prevails today, entirely refounded though it has been on the basis of Mendelian genetics and the concepts of population dynamics — mainly by Sewall Wright, J.B.S. Haldane, R.A. Fisher and, in its more practical aspects, by the research of Theodosius Dobzhansky and his school.

Variation, mutation and candidature for evolution. The candidature for evolutionary change is proffered by the prodigious and virtually inexhaustible range of genetic diversity made possible by the Mendelian processes of segregation and recombination (see Chapter 3). It has sometimes been rather naively objected that the variations so generated are merely variations upon a limited number of themes and that, being essentially of finite compass, they cannot possibly provide material enough for evolutionary change. This objection, however, is like saying that new writing is hardly possible because literature consists only of so many variations upon the twenty-six letters of the alphabet or that Western music must surely be exhausted because it consists of no more than variations upon the notes of the diatonic scale. In reality, prose can be enriched by an enlargement of vocabulary and music by the admission of new notes into the musical vocabulary — e.g. quarter tones. It would thus be reasonable to ask if there were any analogous enrichment of genetical variation which might enlarge still further the candidature for evolution.

New genetical information does indeed come into being in the process known as 'mutation'. Mutations are random perturbations in the genetic material that change the character of the information they convey. They are sometimes sub-divided into those that affect particular genes and those that affect quite extensive lengths of the nucleic acid chain and even of chromosomes. All mutations have it in common, however, that their informational content cannot be *specifically* directed by any environmental event; thus no mutation arises to fulfil or meet any real or imagined need of the organism. All that may be said is that the *frequency* of mutations is increased by certain environmental influences, notably ionizing radiations which can directly or indirectly affect DNA, just as the frequency of occurrence of say the number ten at roulette may increase by throwing the roulette ball more frequently, though ten will not come up more often because it makes or mars the fortune of any competitor. In spite of very strenuous and not always very scrupulous attempts to unseat it, nothing has yet occurred to challenge the belief that the environment cannot act 'instructively', i.e. cannot imprint specific genetic information upon the genetic system of living organisms. No one who really understands the subtlety and enormous explanatory power of the English language need wonder about the adequacy of mutation, crossing over, segregation and recombination to provide rich material for the evolutionary process. Nevertheless it would be foolhardy to say that we knew all there was to know on the subject and that no new source of genetic information will ever be discovered. Generalizations of this degree of dogmatic confidence are almost invariably refuted by the progress of science.

However, the notion that genetic information is self-generated pervades the whole of modern biology and underlies our interpretation of, for example, both antibody formation and bacterial adaptation (q.v.). Some philosophers* generalize the notion still further, and extend the same principle to creative acts of mind.

Natural selection. Charles Darwin was perfectly aware of the animistic connotation of the term 'natural selection', but of course he did not believe that nature was actually selecting anything in the sense in which human beings pick and choose between alternatives. He made it perfectly clear in his correspondence, especially with Asa Gray, that he used the term natural selection in order to avoid the tedious periphrases that would be necessary if on each occasion he used it he were to cast the expression into a suitably objective form. An objective explanation of the principle of natural selection might go as follows (it is long-winded but formally exact).

All human beings alive a hundred years from now will be the descendants of the human beings alive today, so that the human beings alive today make up one hundred per cent of the ancestry of future generations. Human beings, however, are of very many different genetic kinds and it is not to be expected that each genetic kind will make an equal or numerically proportionate contribution to the ancestry of the future population. Some genetic types will take a disproportionately large share and these are accordingly said to be 'selected' and to confer extra *fitness* upon their possessors. The judgment is of course an entirely retrospective one so that the equation of natural selection to the survival of the fittest is a tautologous judgment. It is the pith of the theory, however, that the organisms with the higher net reproduction rate are more highly adapted to their environments than their less favoured contemporaries. The word 'net' in the expression 'net reproduction rate' is of special significance. Even very senior biologists who ought to have known better have complained that modern Darwinism envisages natural

---

* D.H. Campbell in *Studies in the Philosophy of Biology*, eds F.J. Ayala and T. Dobzhansky (Macmillan, London, 1974).

selection's working entirely through the numbers of offspring produced, whereas in reality selective values are expressed by a figure that represents the *net* likelihood of living and reproducing, i.e. chances of an organism's giving birth to offspring that will survive to the age of their parents when they were born (see Chapter 8, Demography).

In the process of natural selection what is or is not selected is an individual organism, but in the classical formulations of the genetical theory of selection the subjects of selection are envisaged as individual genes. In a freely interbreeding population the Mendelian processes do not affect the frequency of individual genes. Gene frequency can be assumed to remain constant from generation to generation except — speaking in a deliberately figurative way — in so far as some external 'force' causes the gene frequency to alter. One such force is 'mutation pressure', which causes the frequency of a mutant gene to increase by reason of the fact that gene mutation is a recurrent process, so that the mutant gene is constantly reintroduced into the population. A second agency which may radically alter the proportion of genes in a population is — as Sewall Wright was the first to foresee clearly — *luck*. In small populations particularly, one cannot assume that the genes present in the gametes are an exactly fair sample of the genes present in the parental population: some genes will increase in frequency by luck, therefore, and others will decline or perhaps even disappear — a phenomenon known as 'genetic drift'. Whatever effects may be attributed to these other agencies, however, all students of evolution agree that the factor which outweighs all others in bringing about a change in gene frequency is natural selection in the sense described above, i.e. the net reproductive advantage of the possessors of some genes over the possessors of their alternative or allelic forms.

The measurement of natural selection in terms of net reproductive advantage is purely and simply a method of *pricing* genetic goods; it is not a statement about their nature or quality.

Representation of evolutionary changes in terms of gene frequencies is not nearly as abstract as may appear at first sight, for a simple algebraic theorem — the 'Hardy-Weinberg

Theorem' — makes it possible to translate any statement about the frequencies of genes in a population into a statement about the frequency with which they will appear in the homozygous or heterozygous forms (see p.33). Thus we can pass at will from gene frequencies to statements about the frequencies of whole genetic make-ups.

The Hardy-Weinberg Theorem.   G.H. Hardy (1877—1947) was the foremost English mathematician of his day and in many ways a rather Olympian figure. He was interested in genetics and soon saw that the so-called 'laws' of inheritance lent themselves to the formulation of a kind of Mendelian algebra, in which the Hardy-Weinberg Theorem is the most important proposition.   It sounds at first like a dull quantification of the apportionment of genes in heredity, but it is in fact of the utmost importance for population genetics, eugenics and almost every real-life genetic context.

The Theorem takes its simplest form when we consider alternative genes which for illustrative purposes we shall call A and *a*.   In so far as these are alternative genes three genetic make-ups are possible: the two homozygotes AA and *aa* and the heterozygote A*a*.   Suppose that the frequency of A is p (say 0.6) and that the frequency q of its opposite number *a* necessarily (1 — p) is 0.4.   The Hardy-Weinberg Theorem states that in a freely interbreeding population, equally divided as to sex, in which the frequencies of alternative genes are as stated above, then the frequencies with which the three possible combinations AA, A*a*, *aa* will turn up in the population will be as

$$p^2 : 2pq : q^2$$

The great importance of this equation is that it makes it possible to translate general and rather abstract sounding statements about gene frequencies in a population into statements about the occurrence of real genotypes.   The importance of the Theorem in population genetics and eugenics will become clear in Chapter 7.

There is an almost Newtonian flavour about the way in which natural selection enters the genetical theory of evolution by selection: the frequency of genes in a population remains constant generation by generation except in so far as some 'impressed force' is brought to bear upon it that causes it to change, and by far the most important of these impressed forces is that of natural selection. Although the usage of 'force' is figurative, natural selection may be said to have both a magnitude and a direction. Magnitude is measured in terms of net reproductive advantage and direction in terms of the nature of the replacement of one specified allele by another.

It would go beyond the authority of any philosophically literate scientist to say that the theory of evolution just outlined was so firmly established as to be beyond the possibility of question or challenge in the future; on the other hand one should be clear where its real or imagined weaknesses lie. There is no serious doubt about the ability of natural selection to bring about the most radical and far-reaching changes in the genetic make-up of a population: the element of the theory that might raise misgivings is that which has to do with candidature for evolution. Even while recognizing the profusion and degree of detail of the heritable variation made possible by Mendelian process with mutation in the background to add extra combinatory symbols, one may still wonder whether the whole story has been told and whether there may not be some other hitherto unrecognized source of variation: some profane young immunologists have indeed used the letters G O D to stand for Generator of Diversity. It is misgivings of this kind that have led so many people — particularly those (among them George Bernard Shaw) with no qualifications to express any opinion on the matter — to question the authenticity of the Mendelian theory of evolution by selection and to prefer the mystical variant, Lamarckism, which will now be outlined.

Chapter Five

Lamarckism

One of the methodological curiosities of evolution theory is that it is a subject upon which everybody feels entitled to hold an opinion: 'If I have indeed evolved,' the sceptic may ask, 'surely I am entitled to an opinion on how I evolved, and from what.' It is on similar grounds that everyone feels himself entitled to speak with authority on Education, for was he not himself educated? The most popular alternative to the views outlined in the previous chapter is that common-ly known as 'the inheritance of acquired characteristics'.

The views on evolution actually held by Jean Pierre Antoine Baptiste de Monet, le Chevalier de Lamarck, are less relevant to present-day perplexities than those traditionally imputed to him under the name of 'Lamarckism'. At the very heart of Lamarckism lies the belief that specific genetic information can be imprinted upon the organism by its needs or by an influence from the outside. It must at once be said that there is no known or, by modern lights, even conceivable method (see Chapter 12, Molecular Biology and Chapter 9, Development) by which information could flow into nucleic acids from the outside.

There are a number of special incentives for believing in Lamarckism — political and psychological in particular. The political incentive to follow Lamarckism and to discredit Darwinism has often been commented on. Lamarck had some influence on French Revolutionary theorists: if men are born equal and yet end up so very different, it must surely be because an individual's character and capabilities are shaped by upbringing, environment and his own endeavours. This being so, it seems only fair and right that distinctions and capabilities so hardly achieved should be passed on to the next generation. One can scarcely wonder that this theory became doctrinal in the U.S.S.R. On the other hand Darwinism, with its insistence on inborn inequality and a competitive process, seems more in tune with a Tory conception of society. To these political incentives should be added the strong psychological inducement to believe in Lamarckian inheritance; cultural, psychosocial or 'exosomatic' evolution (see Chapter 6) is manifestly Lamarckian in style. Moreover, it is fully in accord with our sense of fair play and the fitness of things that a human being's endeavours and physical distinctions (e.g. the musculature of an athlete) should somehow be propagated to his children. In this view the blacksmith with brawny arms, who is so often called upon to testify on these occasions, can not only teach his children to become blacksmiths but can pass on to them some genetic propensity to develop arms as brawny as his own.

Because of these incentives and the right-seemingness of the whole process, the advocates of Lamarckian theory have often been driven into a state of near-exasperation by the temperate contention that every strictly designed and scrupulously executed experiment intended to appraise Lamarckism has faulted it. This also applies to the last remaining strongholds of an 'instructive' theory of heredity: bacterial training and antibody formation. In bacterial 'training' (the word itself is significant) a bacterial culture is slowly, as it seems, 'taught' to use a new substrate or to become resistant to the action of a new antibiotic. In both cases there is a strong natural inducement to believe that some special molecular configuration of the substrate or of the new antibiotic informs the process of protein synthesis so that enzymes capable of breaking down the substrate or the antibiotic are thenceforward formed. In reality nothing of the kind occurs: the process of bacterial training is one of evolution (see Chapter 11, Microbiology) and turns upon the natural selection of variant forms already endowed with the new capability which ultimately becomes the property of the population as a whole, simply because the organisms that possess it become the prevailing forms. In antibody formation moreover (see Chapter 13, Immunology), there is an equally strong inducement to believe that the antigen informs the synthesis of antibodies and so causes an antibody to be formed which is exactly complementary in structure to the antigen. This view has also been slowly and reluctantly abandoned and 'the new immunology', as it has been called, really dates from the recognition, based on the advocacy of Jerne, Burnet, Monod and Lederberg, that the antigen also brings out a pre-existing capability in the antibody-forming cells. Thus 'instructive' theories of metabolic performances no longer have a foothold in biology.

A further and less obvious inducement to believe, contrary to all evidence, that 'there is something in Lamarckism' is the frequency with which it turns out that some adaptation which *could* have been, and looks as if it *had* been, produced by the direct action of the environment is in reality 'laid on' by development, i.e. has become part of the genetic programme — a genetic imitation of an environmentally-produced effect, the 'Baldwin effect'. A case in point is the specially thick skin on the soles and heels of our feet, and the flexure lines on the palms of the hands. In all such cases we feel pretty sure that if the adaptation had not been programmed it would have been produced in an individual's own lifetime merely as a result of use or abuse — in the one case, of course, the habit of walking on the soles of the feet might be expected to produce a thickening of the skin akin to that which occurs in the formation of corns or callosities, and in the other the repeated flexion of the hands at their joints produces lines like those on the faces of people who habitually smile or frown. While admitting that adaptations of this kind offer a special inducement to believe in a Lamarckian theory of heredity it must at the same time be conceded that a great variety of adaptations could not possibly have arisen in this way: e.g. the skin over the front of the eye — the cornea — could not have acquired its special toughness and almost perfect transparency as a result of attempts to see through it, i.e. as an effect of use or of need to use. The tendency of genetic mechanisms to, as it were, take over from the environment is referred to as the 'Baldwin effect' or as 'genetic assimilation' — a phenomenon which has been brought about in a number of ingeniously designed experiments by C.H. Waddington. I am obliged to Sir Karl Popper for a reformulation of the Baldwin effect in terms which satisfy our perfectly understandable predilection for believing that a human being's own wishes, exertions and endeavours can influence his heredity: everything a human being does in ordering his life, in creating or changing the pattern of social institutions, alters his environment and therefore the pattern of selective forces that normally work upon him. The practice of civilized and co-operative behaviour might indeed become part of the genetic programme if matters were so ordered that people with predatory, aggressive and ruthlessly self-interested behaviour were somehow disadvantaged by a society in which mutual support and the habit of co-operation prevailed.

But in spite of all these various inducements to believe in a Lamarckian theory, 'psychosocial' evolution — the subject of the next chapter — is the only context in which Lamarckism holds good; as an agency in ordinary organic evolution Lamarckism is quite discredited.

Chapter Six

Exosomatic (Psychosocial) Evolution

Everybody has observed that the human artefacts which serve as tools are to some extent extensions of the body. The use of the microscope and telescope endows human beings with super eyes and aircraft endow him with the power of flight. Clothing has some of the protective qualities of fur and antibiotics sometimes do what antibodies can not. Geiger counters give human beings a sense organ they would otherwise lack: one that responds to X and gamma radiation. To describe such instruments as 'external organs' or, as Lotka described them, 'exosomatic' organs of mankind is not merely whimsical, because all sensory instruments report back through our ordinary or endosomatic senses and all motor instruments are programmed by ourselves either in their actual operation or by reason of the design characteristics that have been built into them.

It is very clear that these exosomatic parts of ourselves undergo a slow, systematic, secular change of a kind which it is perfectly proper to describe as an 'evolution' — *exosomatic* evolution — provided of course one realizes that it is the design of these instruments that undergoes the evolutionary change and not the instruments themselves, except in a quite unnecessarily figurative sense. The parallels between exosomatic and ordinary endosomatic evolution are amusing rather than instructive. For example, vestigial organs — like the long-since functionless buttons that tailors still put on the sleeves of men's jackets — are to be found in both situations.

Another example, analogous to sleeve buttons, is the tendency of automobile manufacturers to incorporate into their latest models a vestigial remnant of some design that distinguished an earlier *marque*.   A more serious parallel — and one that may be characteristic of innovative changes generally — is to be found in the fact that when evolutionary changes take place in external organs like bicycles and automobiles the changes do not occur concurrently throughout the whole population, but appear first in a limited number of members of the population and then spread throughout the whole, until, because of their economic fitness, they become the common and prevailing pattern. The selective forces in this parallel to endosomatic evolution are, of course, the forces of the market place. Even so, the parallel with the evolution of a population of organisms is obvious enough.

Differences between exosomatic and endosomatic evolution.   Whereas the similarities between these two modalities of evolution range from the amusing to the instructive the differences between them are profoundly important.

1. Ordinary organic evolution is mediated through a genetic mechanism but exosomatic evolution is made possible by the transfer of information from one generation to the next through *non*-genetic channels. By far the most important of these non-genetic agencies is language, surely the most important and distinctive of all human possessions.* It is perhaps because the subtlety, versatility and information-carrying capacity of language is even greater than that of the genetic mechanism that exosomatic evolution is a much more rapid and powerful agency of change, anyhow in human populations, than ordinary organic evolution. It is because of the primacy of language as the agency which provides the link between one generation and the next that exosomatic evolution is often referred to as 'cultural' or 'psychosocial' evolution.   These terms are less satisfactory than the non-committal 'exosomatic' or 'exogenetic', because 'cultural evolution' might so easily be taken to be an evolution *of* culture as opposed

---

* See George Steiner, 'The Language Animal' in *Extraterritorial: papers on literature and the language revolution* (Penguin, Harmondsworth, 1975), pp.66–109.

to an evolution of which culture was an agency.

2. Exosomatic evolution is Lamarckian in procedure: children of mountain-dwelling folk are not born with one leg slightly longer than the other, which would be Lamarckism in the sense now discredited, but if parents tell their children how best to go round the hills the same end is achieved by exogenetic means. Clearly the continued existence of civilization depends on the propagation from generation to generation not just of knowledge and knowhow but also of works of art and all other products of the mind and spirit, together with whatever may be learnt of the wisdom of living.

3. If a box is divided by a partition perforated by a hole just large enough to allow one molecule to pass through, we can amuse ourselves by thinking it possible that all the molecules might accumulate at one end of the box leaving an empty space at the other; going on to reflect that all that prevents its being an everyday occurrence is its very extreme unlikelihood. In just the same spirit we can entertain the idea of ordinary endosomatic evolution's being reversible. All we need to assume is a reversal of the sign of all the selective forces that have hitherto been at work and a reversal or annulment of the mutations and the genetic recombinations that had produced just the right candidature for the retrogressive evolutionary change. This combination of circumstances may also be regarded as improbable to a comical degree. Exosomatic evolution, by contrast, is quite easily reversible, for all that need happen is the total interruption of the cultural links between one generation and the next: not merely a burning of the books, but a disappearance of all human artefacts including all evidence of human cultural institutions. Some of the more infamous tyrannies to which human beings have fallen victim in the past few decades took a few hesitant steps in this direction. The enormity of genocide, it may be added, lies no less in destroying distinctive cultural traditions than in destroying what can never be replaced, a distinctive configuration of genetic endowments. It is hardly possible not to exult in the strength and resilience of a human spirit which has

endured and survived the terrible insults to which it has
been exposed twice in the present century. It is clear that
the bonds between generations established by parental
care and indoctrination are enormously strong.   The
reversibility of exosomatic evolution shows that the
possibility of a return to total cultural barbarism is not
to be dismissed merely as the flight of a morbid imagination, even though it is the fate which each political party
warns us will be the inevitable consequence of voting for
the other.*

Popper's Third World.   Popper's Third World as he has
outlined it in *Objective Knowledge: An Evolutionary
Approach* (Clarendon Press, Oxford, 1972) is a convenient
designation for that which is transmitted from generation to
generation in the process of exogenetic heredity or psychosocial evolution.   The Third World is the objective world of
actual or possible objects of thought, in which may perhaps
be included those material objects which are physical
embodiments of human design; thus a mechanical invention
is a sort of solid hypothesis, of which the adequate mechanical working is the empirical test.

Exosomatic evolution is the great evolutionary innovation
of mankind — the process to which we owe our present
biological supremacy and our hope of future progress.

* P.B. Medawar, 'Technology and Evolution', *The Frank Nelson Doubleday
  Lectures — 1972/73* (Doubleday & Co. Inc., New York).

Chapter Seven

Eugenics

Eugenics is the political arm of genetics. The word was coined by Francis Galton (1822–1911) who introduced it in the following terms:

> Eugenics is the science which deals with all the influences that improve the inborn qualities of a race; also with those that develop them to the utmost advantage.
> Man is gifted with pity and other kindly feelings; he has also the power of preventing many kinds of suffering. I conceive it to fall well within his province to replace natural selection by other processes that are more merciful and not less effective. That is precisely the aim of eugenics.

These sentiments sound dignified, reasonable and humane, but closer inspection of Galton's writing reveals a very sinister streak. He sneers loftily at the efforts at self-improvement made by people lacking the genetic endowments for leadership and, when proposing that the human race could with advantage be propagated through those most richly endowed by nature (as opposed to upbringing), says explicitly that if the less well endowed were to persist obstinately with the perpetuation of their kind, they would forfeit their claim to kindly treatment.

These passages make sorry reading and are largely responsible for the discredit into which eugenics has fallen today. That part of eugenics which insists unconditionally on the primacy of the genetical composition in making people what they are or are not has equally clear political affiliations, for it is a canon of high Tory philosophy that a man's breeding determines absolutely his capabilities, his destiny and his deserts; and it is equally a canon of Marxism that inasmuch as men are born equal a man is what his environment and upbringing can make of him.

As both views are biologically mistaken, it is not surprising that biologists tend to walk delicately when entering discussions on eugenics.

This is a pity, for many genuine genetic dangers go unrecognized because biologists have been reluctant to express their views about the way things are going. Three questions are particularly important and call for draft answers without delay:

1. Do advances in medicine and public health necessarily lead to genetic deterioration?
2. Can anything be done to diminish the burden or the threat of the so-called hereditary diseases?
3. Is it possible to improve the genetic make-up of mankind by procedures that are politically and morally acceptable within the framework of an open society? — i.e. a society in which human diversity is allowed to flourish in all its exuberant variety and in which dissent and disputation, so far from being suppressed, are recognized as agencies of political progress.

The answer to this third question — that which embodies a programme of so-called 'positive eugenics' — is quite simply No: such a project is neither genetically *nor* politically feasible.* We shall therefore consider it no further.

The first question is one of the utmost political and economic importance for any country with a national health service, i.e. one in which the cost of maintaining health is a charge upon the whole population, including the fit.

The answer depends upon whether or not there is any genetic element in differences of susceptibility to disease or the likelihood of cure. That such a genetic element exists is

* See 'The Genetic Improvement of Man' in P.B. Medawar, *The Hope of Progress* (Wildwood House, London, 1974).

known to be true of some diseases and not known to be false of any, so we may take it that the first question applies either directly or indirectly to *all* diseases. The answer to it raises grave politico-economic problems. It is clear that the preservation of relatively unfit genotypes will tend to increase or at least not to diminish their representation in the population. Thus if diabetics are to be kept alive and restored by medical procedures to something approaching a state of normal health, as it is right that they should be, then whatever elements of their genetic make-up may have contributed to their diseased state will for that reason be disseminated more widely throughout the population.

Nevertheless, humanity and self-interest alike oblige us to do our best to meliorate diabetes. It should not be necessary to make such an avowal and indeed it would not have been necessary were it not for a recent revival of a kind of medical Luddism that denounces the entire apparatus of modern therapeutics as something that dehumanizes us and makes us increasingly dependent on the Machine. Moreover some people think, though very few have the brass to say so publicly, that the treatment of disease is in any case an officious interference with the working of God's will or of natural selection. Another insidious argument against a national health service is symptomatic of the 'punishment theory of illness', which is still quite widely prevalent in parts of the United States: if illness is a punishment of sin then a national health service is a deeply irreligious national campaign to prevent the expiation of sin.

These arguments need not be considered in detail. Instead, we should ask ourselves which procedure would dehumanize us more quickly, dependence on the apparatus of therapeutics or indifference to human suffering?

It is nevertheless important to be fully aware of the economic implications of decreasing the load of deleterious genes already borne by the human population. What is happening is that a genetic burden is being translated into a heavy economic burden which will become heavier still, but perhaps we should think ourselves fortunate that such a transaction is possible; most people take the view that life is a bargain at any price. Moreover, therapeutic procedures on a national scale need not diminish fitness in the technical sense explained on pp.44—5; their effect is to create for human beings an ambience in which genes and genetic make-ups no longer gravely diminish the fitness of their possessors in the sense of their ability to contribute to the ancestry of future populations. Unfortunately the environment has to be paid for and is very costly.

We can be fairly confident that the answer to the question 'Do advances in medicine and public health *necessarily* lead to genetic deterioration?' is 'No, not *necessarily*.' Historical accounts of plagues and other murderous pandemics of infectious disease combined with even the briefest appraisal of the bills of mortality of London or New York in the middle of the nineteenth century confirm J.B.S. Haldane's belief that infectious disease is the most powerful selective force that has ever acted upon mankind. It is made clear below, however, that a measure of inborn resistance to infectious disease has sometimes been purchased by genetic tricks or metabolic quirks which are positively disadvantageous in an environment in which people are not at risk of the disease. In such situations it is positively disadvantageous to be genetically forearmed against a danger that does not exist. Indeed it is now clear, thanks to the work originally of Haldane and subsequently of Allison and others, that some forms of genetical abnormality persist in the population because of the' protection they confer against infectious disease. Thus there is now little doubt that the prevalence of sickle cell trait and sickle cell anaemia in West Africa and of Cooley's anaemia (thalassaemia major) and thalassaemia minor in the Mediterranean basin is to be attributed to the modest degree of protection they confer against malaria.

The case of sickle cell anaemia is particularly instructive. When the gene that transforms normal haemoglobin into the abnormal variant haemoglobin S has been inherited from one parent, the heterozygote so formed suffers the mild disability known as 'sickle cell trait', so called because under conditions of oxygen deprivation the red cells collapse into a crescentic or sickle shape. If this were the only genetic cost of securing some degree of immunity to malaria, it would be a price well worth paying, but in reality the medical genetic burden is greater: when two victims of sickle cell trait bear children, then according to Mendelian rules approximately one quarter will be normal, half will be carriers like their parents and one quarter will be the homozygous victims of sickle cell anaemia, which is almost invariably fatal. Nevertheless the heterozygote − the victim of sickle cell trait − enjoys an appreciable measure of resistance to malaria: because this more than outweighs the wastage due to the death of homozygotes with sickle cell anaemia the malignant gene is perpetuated.

It is not easy to draw a moral from this unhallowed balance sheet, but if one is to be drawn it is that if malaria were to be eradicated haemoglobin S and the gene associated with it would slowly disappear − a process which seems now to be taking place in the southern states of America. If sickle cell haemoglobin disappears, so also will sickle cell anaemia. Thus the eradication of malaria will ultimately lead to a genetic improvement, so it does not *necessarily* follow that improvements in medicine and public health lead to genetic deterioration.

This case may not be typical, but in general there is no reason to believe that inborn resistance to infectious disease is advantageous or praiseworthy in any environment except that in which the disease is prevalent. To be genetically forearmed against hazards which do not, or no longer, exist can be likened to the situation of a householder who beggars himself by insuring heavily against risks to which he is not exposed.

Negative eugenics. We turn now to the second of the three questions posed above: can anything be done to diminish the burden or the threat of the so-called hereditary diseases? 'Positive eugenics' may be said to have had the ambition of raising a superior kind of human being − an ambition frustrated both by our ignorance in precise genetic terms of what the outcome of such a procedure should be and of our ignorance of how to achieve it.

'Negative eugenics', by contrast, has the altogether lesser and more realistic ambition of diminishing, and as far as possible correcting, the distress caused by deleterious genes and genetic conjunctions. Even here, alas, ignorance has linked arms with a muddled benevolence to make some foolish and inhumane proposals. One such proposal, of Scandinavian origin, was at one time given some prominence by a society in England founded to promote eugenics: mentally deficient people should be sterilized so as to bring to an end the propagation of the gene responsible for their unhappy state. On closer enquiry, it turns out that the argument was based on the assumption that mental deficiency was due to the conjunction of a single pair of recessive genes which for the sake of argument we will call d, to distinguish it from the normal gene D, so that the genotype of the mentally defective would be dd. The Hardy-Weinberg Theorem described on p.46 then shows how ill-judged such a recommendation is, for if people mentally deficient for this particular reason were to occur with a frequency of 1 in 10,000 in the population then the frequency of the gene assumed to be responsible for the condition would be 1/100 and the frequency of its genetic carriers, the heterozygotes Dd, as high as 1/50. It would indeed be a 'night of the long knives' if it were decided to sterilize one fiftieth of the population, even supposing it were possible to identify the carrier state so as to make sure the knife was wielded on the right people. Only a minority of the offending genes are locked up in the mentally deficient themselves, so sterilizing *them* would not be effective.

In spite of these follies, we emphasize that rationally founded and humane procedures in the area of negative eugenics *are* possible and it is worthwhile outlining the form that some of them might take.

Chromosomal aberrations, such as those which give rise to Down's syndrome ('Mongolism'), Klinefelter's syndrome and Turner's syndrome, being the results of chromosomal accidents of unknown origin are genetically irremediable. Down's syndrome is due to the presence in triplicate of a chromosome which should by rights be present only in duplicate. Its frequency rises so steeply with the increasing age of the mother at childbirth that its frequency in the population generally could be diminished — as indeed it is now being diminished — by a change of fashion or social attitude that favoured young motherhood. When an older

woman conceives a child her physician will often recommend
the procedure of amniocentesis, in which a drop of amniotic
fluid is withdrawn so that the chromosomal make-up of fetal
cells can be ascertained. If the fetus suffers from a gross
chromosomal abnormality the mother might well choose to
have an abortion rather than give birth, even to a perhaps
much longed-for child that was grossly defective.   The
parents' decision will clearly be influenced by their means,
family situation, style of life and religious beliefs, so there
can be no glib formulation of some one correct procedure
that will apply to all cases.

Dominant disorders.   A dominant disorder is apparent
even when the gene responsible for it has been inherited from
only one parent: it is thus apparent in the heterozygous as
well as in the homozygous state. When such disorders are
lethal or very gravely deleterious early in life they are genetically self-correcting, and their frequency in the population
will be determined solely by the rate at which the malignant
gene is reintroduced into the population by recurrent mutation. Nevertheless, specially unhappy problems arise when
the gene exerts its action in middle or later life, for if
children have already been born then natural selection is
powerless to eliminate the gene. Mutations apart, most of
the people who contract one of these diseases — including
Huntington's chorea and a form of intestinal polyposis leading to cancer — will be the children of heterozygotes who
became afflicted *after* they have had their children. Taking
now D as the symbol of the malignant gene and d as its
normal counterpart, the afflicted will generally be of the
composition Dd. Approximately half the children of the
afflicted will therefore contract the disease and all will live
in continual dread of doing so. Under these circumstances
charity suggests that potential carriers should decide not to
become parents. Much the same is true, though for slightly
different genetic reasons, of suspected carriers of the sex-
linked recessive gene responsible for haemophilia, in which
blood clotting is gravely impaired. In general, it is males who
are overtly afflicted and females who are carriers.   On
average, half the sons of a female carrier will be bleeders and
half her daughters will be carriers. In addition to all the more

obvious hazards — even minor operations such as tooth extraction are very dangerous — bleeders may suffer great pain as a result of bleeding into the joints. A great burden of misery would be lifted from mankind if women whose family history suggested they might be carriers of the haemophilia gene were voluntarily to abjure motherhood — though such a decision would have played havoc with the royal families of Europe, as J.B.S. Haldane made clear in a brilliant essay entitled 'Blood Royal', published in the *Daily Worker* in the early 1930s.

Recessive disorders do not make their appearance unless the gene responsible for them has been inherited from both parents. In the small but rapidly increasing proportion of such disorders in which the carrier state can be identified they are eugenically manageable. Although the disorders are individually rather rare, they form between them a numerous and important category, including certain grave inborn defects in the power to metabolize foodstuffs (phenyl-ketonuria and galactosaemia). Most of the overtly afflicted are the children of parents who are both carriers: approximately one quarter of their children will be afflicted and half will be carriers too. To illustrate the possible eugenic management of such conditions along the lines first suggested by J.B.S. Haldane, we shall consider a 'recessive disease' with a frequency in the population of 1 in 40,000. In such a case the frequency of the gene responsible for it will be 1/200 and the frequency of the carrier state 1/100. Eugenic management turns on the possibility of being able to identify the carrier state. Most overt cases of the disease could be eliminated in one generation if, having been identified, the carriers of the *same* harmful recessive gene were to be discouraged from marrying *each other*, or at least from having children by each other. The one carrier among each 100 people would therefore have to find a mate from one of the other 99, among whom only a few would be disqualified by carrying some other unfavourable recessive gene also possessed by the intended spouse.

What is being proposed here is that carriers of the *same* harmful recessive gene, when they can be identified, should either be discouraged from childbearing or warned of the consequences of doing so — to wit that approximately one quarter of their children will be afflicted by the malady of which the gene is a determinant. This proposal has been so far misinterpreted by an eminent anthropologist as to impute to us the recommendation that all carriers of harmful recessive genes should be discouraged from marrying or childbearing — a prohibition which would inevitably bring human reproduction to a standstill, for we all carry several harmful recessive genes. People temperamentally inclined to believe themselves the victims of conspiracy, particularly one engineered by scientists, will doubtless regard this whole proposal as a gross invasion of human rights to be resisted tooth and nail. There is not much likelihood, however, that engaged couples would be afflicted by the fear of being victims of a conspiracy. It is well known that in the past engaged couples sometimes brought their engagements to an end because the prospective groom's blood was rhesus positive and the bride's rhesus negative — an excess of fearfulness on their part because haemolytic disease in the newborn is not an inevitable consequence of such a union: rhesus incompatibility is a necessary, but not a sufficient condition. In any event, it can be prevented.

The Beethoven fallacy. We should like to discuss here the rights and wrongs of an argument against any form of radical eugenics that involves the exercise of the power to terminate pregnancy. It owes its name to the particular form in which the argument was put by Mr Maurice Baring and it was recounted by Mr Norman St John Stevas, M.P. in the following terms:

> One doctor to another: 'About the terminating of pregnancy, I want your opinion. The father was syphilitic. The mother tuberculous. Of the four children born, the first was blind, the second died, the third was deaf and dumb, the fourth was also tuberculous. What would you have done?' 'I would have ended the pregnancy.' 'Then you would have murdered Beethoven.'*

* From *Life or Death: Ethics and Options*, ed. D.H. Labby (University of Washington Press, Washington, 1970).

The reasoning involved in this odious little argument is breathtakingly fallacious, for unless it is being suggested that there is some causal connection between having a tubercular mother and a syphylitic father and giving birth to a musical genius the world is no more likely to be deprived of a Beethoven by abortion than by chaste abstention from intercourse or even by a woman's having a menstrual period; for by either means the world may be deprived of whatever genetic make-up conduces to the development of a musical genius.

In summary, the problems of eugenics are genuine, in urgent need of serious consideration and, where possible, of solution. However, any solution that is proposed or attempted must give due weight to political, economic, humanitarian and religious considerations.

Chapter Eight
Demography

In the United Kingdom patronage of demography is claimed by the British Academy — and therefore in a sense by the humanities — rather than by the Royal Society and the natural sciences. It may be regarded either as the most 'scientific' of all the behavioural sciences or alternatively as that branch of applied mathematics which overlaps most extensively with sociology. It has already been explained that certain fundamental demographic ideas are an important part of the genetical theory of evolution by natural selection. Demography has to do with the structure and growth of populations. No natural population is adequately characterized merely by its number, for it has a structure both in space — a characteristic distribution in the environment — and in time — the members of the population will have a certain distinctive distribution of ages — so many of reproductive age, so many below and so many above, etc.

As in any growing system in which the products of reproduction are themselves capable of reproducing, the *norm* of population growth — that theoretical standard from which all real instances of growth have so many departures — is exponential or logarithmic, i.e. the population increases by an equal proportion in each unit of time, as in 'compound interest', rather than by equal increments in each unit of time, which would be merely 'simple interest'. In representing population growth graphically it is customary to plot the *logarithm* of the size (= number) of the population against time. The reason for this is that the logarithmic scale is one in which equal sub-divisions represent equal *multiples* and not, as in the ordinary scale, equal increments: in the very simplest case, therefore — growth at a fixed rate of compound interest (Malthusian growth) — the logarithm of population size forms a straight line when plotted against time. Needless to say no real population can grow by continuous compound interest for more than a relatively short time, and logarithmic growth in real life is realized only for short periods by bacterial populations growing in media from which the waste products are as far as possible continuously removed. In real life the growth of a population is necessarily restrained by one or more density-dependent factors, i.e. by one or more factors whose effect increases as the population increases. It is a matter of some satisfaction to advocates of temperance that in populations of such simple organisms as yeasts, the accumulation of alcohol is one such density-dependent factor. Others are of course shortage of food and lack of *Lebensraum*.

The information the demographer has to work on consists of (a) census data and (b) certifications of births and deaths. It is the latter that make it possible for the demographer to convert the still picture provided by a census into a motion picture which may help him to make limited predictions about the future. Predictions about the likely future size of populations are particularly important nowadays and indeed have been for the last thirty or forty years. For this reason demographers have in the past felt themselves under some pressure to produce a single-value statistic embodying a measure of the population's reproductive vitality — a process somewhat akin to taking the population's temperature to assess its state of health. Among such supposed measures of reproductive vitality are the 'net reproduction ratio' (already mentioned on p.44 in the context of evolution theory) and the closely related Malthusian parameter of A.J. Lotka, borrowed without acknowledgement by R.A. Fisher in his treatise on *The Genetical Theory of Natural Selection*. All such indices are measures of fertility duly weighted by mortality, for it is not sufficient to estimate or predict the chances of a woman's having a female child: we must go beyond this to ask what the chances are of this child's growing up and living to the age her mother was when she bore her. The *net reproduction ratio* is in effect the ratio of live births in successive generations, usually but not necessarily estimated with respect to the female population only. The Malthusian parameter — Lotka's 'true rate of natural increase' — is also a net figure expressing the population's prevailing rate of growth by continuous compound interest.

Nowadays demographers regard all such single-figure measures of a population's reproductive vitality as unduly simplistic. To make a plausible estimate of what the size of a future population is likely to be we need to know its age-distribution, the reproductive rates of its members at each age and of course the age-specific death rate.* The age-specific death rate or 'force of mortality' is the number who die in each age interval (say 60 to 61), expressed as a fraction of those alive at the beginning of the interval, i.e. as a fraction of those who qualify to die in the chosen age interval. Demographers are now altogether disenchanted with the idea of finding some *one* measure of a nation's reproductive vitality: the principles of the much more subtle methods used nowadays to attempt to predict future population sizes are outlined below.

Life tables and cohorts. The compilation of a life table is fundamental to the actuary's craft. The best way to envisage the construction of the life table is to imagine that a thousand or hundred thousand or some other large round number of human beings or other organisms are labelled or otherwise identified at birth and then followed throughout life, recording the ages at which each individual dies, until the death of the last one. The table therefore begins with a figure such as 100,000 and ends at zero. From such a table we can derive the age-specific death rate directly, for it represents the fraction dying during each year or epoch of life of those actually at risk — for example, the number who die between the ages of 40 and 45 expressed as a fraction of those who were alive at the age of 40. This 'force of mortality' is lowest

---

* The crude death rate of a population is the death rate per unit of time per so many members of the population — perhaps the death rate per thousand per annum averaged over the population as a whole. Obviously such an estimate must be very sensitive to changes in the age-distribution of the population.

at ages 13 to 15 — the actuarial prime of life — although it is relatively high around birth and in infancy and of course rises throughout the rest of life.

Another important and informative statistic that may be derived directly from the life table is the *mean expectation of life.* One can ask of newborn children, or indeed of people of any age, 'How much longer, on the average, may a person of this age expect to live?' The mean expectation of life at birth is a figure widely quoted for those countries (a minority) in which the relevant information is available and it is sometimes used as an index of national wellbeing or national advancement. There is some sense in this provided one realizes that the principal cause of the steep increase in the mean expectation of life at birth in the Western world over the past hundred years has above all else been the decline of infantile and neonatal mortality. Considered as a measurement of something, therefore, the mean expectation of life at birth can be taken as a measure of the degree of success that has been achieved in controlling infectious disease. The connection between demography and population genetics is made obvious enough by reflecting that the genetic composition of a cohort of individuals considered collectively changes during its lifetime, i.e. the genetic make-up of the cohort as it approaches the end of its life could not be a fair sample of that of the starting population when the life table was taken to begin, for there are important inherited differences in vulnerability to mortal hazards of all kinds.

Cohort analysis applied to the study of fertility. The strength of cohort analysis in the study of mortality is that the unit element of the life table is something which has a real meaning biologically: it is a life. Similar considerations apply to the use of the cohort method in studying fertility. Instead of studying the fertility of women 'cross sectionally' at a number of different ages, the method of analysis introduced by D. Glass and J. Hajnal is based upon a study of the reproductive performance throughout life of a cohort of women from the beginning of the age of reproduction to its end. It is thus possible to get an idea of the variations in pattern of family building. For predictive purposes a specially important statistic is *completed family size* and it is in making predictions about the sizes of completed families in years to come that knowledge of the pattern of family building is so important.

The biological principles of population control.   It is an historical accident that the principle of natural selection is most often introduced to young students by means of a syllogism that runs as follows:

1. Organisms produce offspring in numbers vastly in excess of their requirements.
2. Only a small minority of these survive to the age of reproduction.
3. Natural selection must act in such a way as to preserve the organisms best fitted to survive.

The first proposition in this argument embodies a grave fallacy.   Organisms do *not* produce offspring in numbers vastly in excess of their requirements.   Whenever an exact study has been made of the correlation between fertility and mortality, it has turned out that organisms produce just about the right number of offspring to ensure their survival. The most thorough of these studies is that of David Lack on nesting birds.   Nesting birds lay a characteristic number — a 'clutch' — of eggs, and there is every reason to suppose that this number is under the control of natural selection: a smaller number would prejudice the chances of survival, and a larger number of nestlings would put a physiological strain upon the mother which she might be unable to meet.   Any particular clutch size probably represents a compromise between these two factors.   To envisage an unlimited growth of population as in premise 1 above is therefore to combine a real rate of fertility with a quite imaginary rate of mortality. All real populations are limited in size by density-dependent factors, as explained above (p.67).   In human populations a special anxiety is, of course, that these density-dependent factors will include major sources of human distress, such as death from starvation and infectious disease.   It is therefore imperative to restrain the growth of human populations so that they do not become prey to any such disasters.   The present situation, as is well known, is that mortality is in decline, without any proportionate reduction of fertility, so that the very situation envisaged by the Malthusian syllogism outlined above is being realized. Although the principal obstacles to the adoption of birth control procedures are educational and administrative, there are biological obstacles as well: these may be summarized by saying that there is a physiological conspiracy against the adoption of effective birth control procedures. No behaviour has deeper evolutionary roots than reproductive behaviour and the biological 'forces' that do so much to promote fertility and reproduction cannot be easily annulled; nevertheless, the problem is a soluble one and fortunately none of the many scientists trying to solve it are deterred by the foolish accusation that their work might be said to betray an enmity to nature.

In intervals between 'calling for' improvements in our way of life, public figures often call. for the stabilization of the population at whatever number may seem to them to be politically desirable. Such public figures obviously have no idea of the extreme difficulty, indeed — short of tyranny — the demographic impossibility of stabilizing the size of a population at some predetermined number.

It is, however, possible to take measures which will keep a population within bounds, but these measures are of course political, not biological. They take the form of fiscal or educational inducements or deterrents to having larger or smaller families as the case may be, but even then it is notorious that population size is affected by factors as imponderable and unpredictable as fashions for a particular family size or pattern of family building. Such fashions can of course be strongly influenced by political and religious as well as by economic and prudential considerations.

An idea which has gained widespread currency is that if all married couples were to limit the number of their children to two, then the population would be stabilized. This proposal has that air of commonsensical rightness about it which is almost invariably a symptom of some aberration of reasoning. The trouble is that the two-child proposal gives no weight either to mortality or to infertility — factors of the utmost importance for the reproductive vitality of a population.

The story that from time to time colonies of lemmings plunge headlong over the edge of the cliffs to perish in the sea is one that has derived its air of authenticity from the earnestness with which it is recounted, which may be thought to outweigh the fact that everyone who tells the story does so on the authority of someone else. Biologically, the evolution of such a process is almost inconceivable and indeed the evidence that it occurs at all is far from convincing. It is now recognized as part of the mythology of biology.

The evolution of altruistic behaviour raises a number of very difficult problems, but none so difficult as to explain the evolution of a form of altruism such that any genetic factors supposedly responsible for it perish with their possessors in a kind of demographic *auto-da-fé*. In nature, animal numbers are regulated by *force majeure* — starvation and death by predation or infectious disease — and in the long term by the influence of natural selection on fertility. If family limitation were to be practised by humans for dozens of generations, their fertility *might* decline because the hugely fecund would no longer enjoy a selective advantage over the prudent. If such a decline were to recur, we should regard it as the effect of a natural and beneficent evolutionary process that brought fertility into line with our greatly reduced mortality — an evolutionary adaptation comparable with that which has occurred in other species in which fecundity is under the influence of natural selection.

Chapter Nine
Development

In the section on the principles of population control (p.70) grave doubt was expressed about the validity of the major premise of the Malthusian syllogism — that which declares that organisms produce offspring in numbers vastly in excess of their needs. The same doubts apply to the production of germ cells — sperm and eggs: however large the number produced, it is not likely to be greater than that which is just about enough to make good natural causes of wastage, for in the strict sense, all spermatozoa are wasted except those that unite with ova. Ova are not produced in such profusion. The work of Zuckerman and his colleagues showed that contrary to the prevailing opinion ova are not formed anew throughout the life of mammals: an entirely adequate number are already formed by the time of birth and these ripen and are shed periodically throughout life. The child of an older mother therefore develops from an older egg — an egg which has been longer exposed to all influences inimical to eggs. It is therefore not surprising that, as a class, children born of older mothers differ in certain distinctive ways from the children of younger mothers: for one thing, the frequency of that form of Down's syndrome which is due to the triplication of a chromosome that ought only to exist in duplicate is somewhat higher and so also is the frequency of fraternal (as opposed to identical) twinning.

Animals that reproduce sexually all start their life as a fertilized egg, a 'zygote' formed by the union of a single sperm with a single egg. In this process of union the diploid number of chromosomes is restored (see p.30).

The eggs of vertebrate animals are astonishingly diverse

considering that each is, in basic anatomy, only a single cell; in that light the yolk of an ostrich's egg is probably the largest cell to be found in animal or plant kingdoms. There is equal variety in the character and conformation of the various membranes and envelopes that surround the egg, among which can be included the tough almost leathery capsule of the shark's egg and the hard calcareous shells of birds' eggs.

The eggs and embryos of all vertebrate animals develop either actually or effectively in water. Amphibians such as frogs lay their eggs in water, but in the higher land vertebrates — called 'amniotes' accordingly — provision for an aquatic environment is made by a special embryonic membrane, the *amnion,* which provides each embryo with a private pond. The human embryo, like the embryo of all true mammals, develops within the uterus of the mother and derives oxygen and food from the maternal circulation. There is no direct connection between the blood circulations of the fetus and the mother: instead, all physiological transactions take place across a 'placenta', in which the capillaries of the maternal and fetal circulations are very closely juxtaposed. In amniote animals waste products are excreted into a special sac, the 'allantois', which begins as an outgrowth from the hind end of the gut and abuts deeply into the amniotic cavity. Where the blood vessels in the connective tissue outside of the allantois reach the blood vessels in the connective tissue that forms the inner lining of the amnion a double membrane is formed — the 'Chorioallantois'; in higher mammals the placenta is of the kind known as chorioallantoic because it is formed by a very close approximation of the chorioallantoic membrane to the uterine wall. In birds the chorioallantoic membrane lies immediately under the porous shell and is the pathway of all gaseous exchanges.

In growing from a fertilized egg to an adult weighing fifty kilograms the embryo increases several billion-fold in size. This increase, which is all that is technically implied by the word 'growth', is the least interesting aspect of development which includes also *morphogenesis*, the taking of shape, and *differentiation*, the process by which all tissues and cells of the body come to acquire their special and distinctive characteristics. All development begins with the internal subdivision of the fertilized egg into separate cells — segmentation or 'cleavage' — and in chordate animals (Chapter 10) this leads to the formation of a hollow sphere of cells, the so-called 'blastula' or something morphologically equivalent to it. In the days when people believed in a naive form of the theory of recapitulation (a notion discussed below), it was supposed that this blastula represented an early stage in the evolution of animals generally, but no such notion is credited today. In eggs containing very little yolk, cleavage is total, i.e. the entire zygote participates in it, but where the yolk is abundant and the egg large, cleavage is partial and confined to a small area — the germinal disc — at one pole of the egg. The cleavage of human and other mammalian eggs is total because the mammalian egg is yolkless, although much of mammalian development still retains the indelible stamp of a reptilian origin. Shortly after the formation of a blastula or something equivalent to it, all chordate embryos undergo a profound metamorphosis known as 'gastrulation', an intricate and finely concerted pattern of cell movements and changes of cellular shape that complete the formation of a rudiment — the 'archenteron' — of what is to be the adult gut. Much significance used to be attached to the fact that at this stage in the development of reptiles and birds the embryo is organized into a set of cellular layers or cellular sheets. These so-called 'germ layers' were at one time pillars of comparative anatomy, for it was believed that each germ layer necessarily gave rise to a characteristic family of cells or tissues in the adult and any cell or tissue that appeared to be derived from an atypical germ layer was thought hardly respectable. Nowadays it is easy to see that the organization of the embryo into layers, tubes and sheets is only a structural reflection of the fact that the tactics of morphogenetic processes consist essentially of the movement of cells and cell sheets relative to each other. The form-building processes which would be impossible with a solid ball of contiguous cells are mechanically feasible when the embryo is organized into layers, tubes and sheets.

The next major metamorphoses are those that lead to the

establishment of the typical chordate form — 'neurulation'. In neurulation an influence emanating from the roof of the primitive gut causes the outermost layer of the rudimentary skin of the embryo to fold inwards along the dorsal midline to form a hollow tube that closes from front to rear and runs from end to end of the body. This tube is the embryonic rudiment of the central nervous system and from the mechanics of its formation it has a posterior opening which closes over, except in cases of the developmental abnormality *Spina bifida*. Under the nerve tube runs the stiffish rod-like notochord which gives its name to the entire group.

The brain begins as three swellings at the front end of the nerve tube: fore-, mid- and hind-brains. Much of the development of organs follows the same plan as 'glassblowing operations' as Oliver Wendell Holmes *Snr* called them: thus the rudiment of the eye begins as an outpushing from the forebrain which caves in from the outside inwards. This rudiment is the so-called eye-cup. A chemical influence emanating from it causes the very delicate embryonic skin that overlies it to thicken to form a lens. More will be said of such 'inductive' processes later. Chordates are segmented animals: the main body musculature begins in the form of a series of blocks of muscle lying on each side of the notochord; these blocks of muscle occupy the body from end to end though the most anterior become modified to form the muscles that rotate the eyeball. Because the muscles are segmented, so are the motor nerves that supply them and so also are the sensory nerves that run between them, but the primary segmentation is that of the musculature.

At the enlarged anterior end of the primitive gut — the 'pharynx' — vertical slits form — branchial clefts — which are anatomically equivalent to gill clefts and are responsible for the idea that human beings had gills at some stage in their life history (see p.80, *Recapitulation*). Skeletal elements support the tissue between these clefts and in the various vertebrate animals that have jaws ('gnathostomes') this skeletal matter is used to form the suspensory system of the jaw. In the highest vertebrates the lower jaw is reduced to a single bone, the dentary, but the remaining bones become transformed into the little ossicles that communicate sound vibrations from the ear drum to the organ of hearing.

Embryonic induction. Mention has already been made of the fact that the formation of the central nervous system (the neural tube) and the lens of the eye depend upon an influence emanating from the roof of the primordial gut and the eye-cup respectively. This process of 'induction' was discovered by the great German embryologist Hans Spemann. The validity of the results which he himself obtained by the study of amphibian embryos was extended by Needham and Waddington to the embryos of amniote animals. Spemann attached very special significance to the region (in Amphibia, the dorsal margin) of the tissue that invaginates in the process of gastrulation, to which he gave the name 'the organizer'. Many later embryologists tend to question the notion of a single organizational centre in the young embryo, mainly because of the reorientation of thought brought about by inwardly digesting the lessons for embryological theory of molecular biology. The most important of these lessons is that the process of diversification in the cells of embryos must represent a different kind of awakening in each cell of some genetic potentiality already contained within it, for it is known that cell division in development is symmetrical, so that all cells in the body start with the same outfit of nuclear genes.

The teleology of inductive relationships is best understood if we imagine what the consequences would be if that tiny little patch of the embryo's outermost layer of cells, which eventually becomes the lens of the eye, were rigidly predetermined at an early stage to do so. If this were so, it is hardly possible that after all the complicated cellular movements of early development this tiny little patch would always end up in exactly the right place opposite the eye-cup which, as already explained, is an outpushing from the forebrain. Inasmuch as something emanating from the eye-cup induces lens formation in the overlying embryonic skin, this particular source of error is eliminated: the lens will form where it fits. Inductive relationships are naturally of special importance in embryos as various as those of vertebrates. It is these inductive relationships that have made possible the great adaptive radiation of vertebrate embryos.

The decoding of genetic information is discussed in the chapter on molecular biology (q.v.).

All biology students soon come to learn of the venerable old antithesis between 'preformation' and 'epigenesis'. It is quite certainly not true that the embryo is anatomically preformed in either of the two gametes out of which it grows, though it was once thought that a spermatozoon was a minute *homunculus* that had merely to enlarge symmetrically to assume the adult form — the female being regarded, as in backward cultures she still is, as merely the soil for the nourishment of the male 'seed'. Christmas crackers sometimes contain tightly rolled tubes of paper which when immersed in water enlarge and unfold into what used to be called 'Japanese flowers'. There is no preformation in this literal sense, but there *is* some real meaning in the antithesis: the genetic instructions according to which development proceeds are indeed preformed, but their realization is *epigenetic*, i.e. turns upon influences acting upon the embryonic cell from the outside: the Japanese flower is preformed but the immersion in water is the necessary epigenetic stimulus.

Neoteny. The overall rate of development, in so far as it can be measured, and the relative rates of development of the different organ systems are not fixed and immutable but can vary within quite wide limits. One of the most important of these variations is that which takes the form of *neoteny*, or as de Beer preferred, the more general term, *paedomorphosis* — the state of affairs in which an animal becomes fully mature and reproduces itself at a stage of development equivalent to that of a relatively juvenile or even an embryonic form of one of its evolutionary ancestors.

It has long been recognized that the development and sexual maturation of human beings is retarded in relation to, for example, that of chimpanzees. Fetal ape-like characteristics of human beings include the relatively enormous brain and the roomy, domed skull that goes with it. Moreover, the slowing down of the later stages of human development in relation to the ape's is very obvious when comparing the times of eruption of the two dentitions, the times of onset of sexual maturity and the length and tempo of the growth span generally. Adolf Portmann described the newborn human being as an 'extra-uterine fetus': the relatively enormous head of a human fetus at birth is such as to make its passage through the pelvis a pretty strenuous exercise.

The idea of human neoteny was used by Aldous Huxley in his novel *After Many a Summer*. Though not to be taken seriously as literature, it embodies the amusing zoological conceit that if a human life were to be prolonged for many years beyond its natural span the human being would ultimately develop adult ape-like characteristics.

Gavin de Beer, whose thought and writing has done so much to clarify the (until then) confused notion of 'recapitulation', pointed out that the specially helpless state of the human newborn must have played an important part in establishing the nuclear family as a unit of social structure.

By far the most striking example of neoteny — and pretty obviously that which gave Aldous Huxley his idea — is that of the axolotl: like many other salamanders it retains a number of larval characteristics, including external gills, throughout life and breeds in that juvenile state. With some such paedomorphic animals the administration of thyroid hormone allows development to proceed to adult life.

Paedomorphosis in evolution. Paedomorphosis turns out to have been a fundamentally important evolutionary stratagem which has offered escape from the extremities of adaptive specialization and opened up entirely new evolutionary possibilities.

It was believed by W. Garstang that the earliest chordate animals — our own remotest ancestors — arose paedomorphically from animals akin to the tiny free-swimming larvae of sea urchins and sea stars. This hypothesis is quite widely regarded as being too far-fetched and speculative to have been 'received' at, say, textbook level. On the other hand, there is very little disagreement that the true chordates, especially the animals classified as *cephalochordates* — contemporary representatives of animals of the kind that were surely ancestral to all the vertebrates — are paedomorphic variants of the larvae of sea-squirts — often called 'ascidian tadpoles'. Cephalochordates begin their development along lines almost identical to those of sea-squirts and like the embryos of sea-squirts they too undergo a strange metamorphosis in the course of development — one that transforms them from a rather grotesque asymmetry to the symmetry of the adult.

The only other important group comparable to the vertebrates which may have arisen paedomorphically is the insects, but other lesser groups for which a paedomorphic origin has been suspected include the nematodes and odd little aquatic organisms known officially as *Rotifera* but unofficially as 'wheel-animalcules'.   In general, however, such vernacular names, thought by Victorian zoologists to be on everyone's lips, have long since fallen into disuse; the agile little freshwater protozoon *paramecium* is no longer called 'the slipper animalcule', if it ever was so: indeed, the word 'animalcule' itself no longer comes naturally to the lips.

Recapitulation. This is a suitable context to enquire into what element of truth there may be in the famous but gravely misleading (it gravely misled Freud) notion of 'recapitulation'.   In its very simplest form, the law of recapitulation states that in its development an animal necessarily rehearses its own evolutionary history, passing through stages comparable to the adult stages of its various evolutionary forerunners.   Thus a human being, starting as a single cell like a protozoan, passes through a stage like the simplest imaginable many-celled organism and then through the stage of having a two-layered structure, like some of the simple hydroids, and in due course through a fish-like stage, in which he supposedly has gills, until finally he acquires a recognizably human shape.   Most of these similitudes are quite false or grossly exaggerated: at no stage does the human being have gills, though in the embryo the great cavity into which the mouth opens, the pharynx, is indeed perforated by clefts which communicate with the amniotic environment, which, according to recapitulation theory, we are to think of as an evolutionary remnant of the sea.

There is, however, an element of truth in the theory of recapitulation which is two-fold.   In the first place, von Baer's principle is undoubtedly valid — it states that embryos of related organisms resemble each other more closely than do the respective adults into which they develop.

In the course of later development vertebrate animals become more and more unlike, according to their several genetic endowments, but all vertebrate animals pass through a stage of development, the *neurula*, so similar in basic structure in the different vertebrate groups that one has to be quite an expert to tell them apart. However, von Baer's law, although useful and valid, is merely descriptive and it does not go quite far enough. The circumstance that makes sense of the law of recapitulation is this: when one speaks of, for example, 'the evolution of an organ A into an organ B', this is best thought of as a shorthand way of saying 'the evolution of the developmental process that led to the formation of organ A into that which leads to the formation of organ B'. This is not pedantry, but an illuminating formulation which makes it understandable that the embryological processes leading to the formation of B should begin with those that led to the formation of A. In other words, the modified law of recapitulation states that under certain circumstances the development of an advanced organism recapitulates the embryonic stages of its evolutionary ancestors — which is very different from saying that an organism 'climbs up its own family tree'.

The observations made above about germ layers and the theory of recapitulation illustrate how seldom it is in science that a theory is wholly disproved and discredited: if inadequate, its fate is usually to become a theorem in — a special case of — some larger theory. The familiar statement that the History of Science is merely a history of errors is thus seen to be a crude philistinism.

Chapter Ten

Bodily Constituents

The physico-chemical properties that make possible the building up of the huge and complicated molecules which form the substance of living organisms are (a) the ability of carbon atoms to link with each other to build the long carbon chains that form the backbone of proteins and of fats, and (b) *polymerization*, the building up of large — sometimes huge — molecules by the additive combination of building blocks that resemble each other in general chemical structure.

In *proteins* the unit of structure or *monomer* is the amino acid, a quite simple organic molecule of which rather more than twenty different species are now known. The combinational variety of these twenty different amino acids confers upon proteins their prodigious diversity of structure and function. Amino acids are described as *essential* (a relative term) when an organism cannot synthesize them and must therefore be able to recover them ready-made from its diet. The condensation of amino acids to form proteins is not a highly energetic process, and very little energy is liberated when protein is disassembled into its constituent amino acids in the course of digestion — otherwise the stomach and insides generally would get pretty hot in digesting a meal.

Nucleic acids are giant polymers built up of nucleotides, each nucleotide a compound of an organic base, phosphoric acid and a sugar molecule. The character of the sugar molecule divides nucleic acids into two main classes: ribonucleic acid (RNA), in which the sugar element is ribose, and deoxyribonucleic acid (DNA), in which the sugar element is deoxyribose. The organic bases in DNA are of four different kinds. The very great biological importance of the linear ordering of these four nucleotides and the nucleic acid polymers is described in the chapters on inheritance and molecular biology (3 and 12).

Ribose and deoxyribose are both five-carbon sugars — 'pentoses'. The ultimate structural element of giant polysaccharides of plants and animals, such as starch, cellulose and glycogen, is often a simple six-carbon sugar (hexose) such as glucose or fructose (fruit sugar). Considered as the monomers out of which the large polymers are compounded, the simple carbohydrates are compounds containing carbon and the elements of water — i.e. hydrogen and oxygen — often in the proportions in which they occur in water.

Fats and oils are compounds of the general kind known as *esters* formed between glycerol and the higher fatty acids. They have nothing like the variety of proteins or polysaccharides. Many of their physical properties turn upon the length of the carbon chain in the fatty acid that contributes to their formation. Their chemical properties derive principally from those of the fatty acid itself: an important distinction is between those fatty acids (unsaturated) in which the combinational capabilities of the carbon atoms are not fully used up, and those (saturated) in which they are. The 'hardening' of vegetable oils is a process of saturation which serves incidentally to raise their melting points, so that they become solid or semi-solid at room temperature and therefore more easily usable as substitutes for butter.

Compounds of still greater complexity than any described above are built up by combinations of polymers of different kinds — e.g. combinations of fatty matter with protein or with carbohydrate, or combinations of all three. Chromosomes and the nuclei of cells generally consist very largely of *nucleoproteins*, salt-like combinations of DNA with a basic protein such as histone, though the protein itself makes no known contribution to the information-carrying properties of the chromosome. Its function may be to package the nucleic acid neatly in such a form that no casual leakage of DNA can disrupt the genetic or developmental process.

Dynamical state of bodily constituents. One of the most

important discoveries that resulted from the introduction
into biology of the use of isotopes of the elements was that
of *turnover*, i.e. the continual replacement of the elementary
constituents of the body: a principle that applies not merely
to those tissues and tissue ingredients already known to be
undergoing a process of constant regeneration, e.g. the
surface scales of the skin and the internal lining of the gut,
but also to structures like teeth and bone that usually strike
us as fixed and semi-permanent. These, too, participate in a
continuous turnover which makes it clear that what is endur-
ing in the body is its form, or the system of preferred stations
occupied by incoming molecules as they take the place of
those already there. The turnover rate of bodily constituents
varies widely from tissue to tissue; that of a tendon, for
example, is specially slow.

The use of isotopes to uncover this important biological
principle is made possible by the fact that though these
variants of the chemical elements are physico-chemically
distinguishable either by radioactivity or by their differing
atomic weights, they are treated by the body exactly like
their 'ordinary' forms. A still further use of the body's
inability to distinguish the elements from their isotopes lies
in the use of isotopes as labels, tracers or markers in the
complex metabolic processes of the body, for the track of a
molecule during its various transformations may be followed
very exactly by the use of isotopes. In modern biological
laboratories, devices for radioactive counting are as familiar
as microscopes: nearly all biochemical research is now cast
into a form in which quantitative estimation is made to take
the form of radioactive 'counts'. Indeed, it is not an exagger-
ation to say that the use of radioisotopes has been a technological revolution in biology as important as the introduction
of the microscope itself.

Enzymes. In living organisms the molecular transformations that accompany nutrition, the storage and liberation of
energy and the decoding and translation of genetic information are all mediated through the action of enzymes, proteins
of a class that, though they may in general be described as
catalysts, do not merely expedite but often make possible
the chemical transformations they promote. Enzymes often

act in series and thus make possible sequential transformations of great complexity, such as those involved in cellular respiration and the mapping of a DNA nucleotide sequence into a polypeptide chain. The action of enzymes is very exactly modulated by co-factors and sometimes inhibitors. The substance upon which an enzyme acts is known as its 'substrate' and the relationship between an enzyme and its substrate is often very specific. The efficiency of an enzyme usually depends on its working in exactly the right conditions of salinity and acidity or alkalinity.

The control and detailed specification of enzyme synthesis is, of course, the work of DNA and it is probably through enzymes that DNA exercises its developmental effects, for enzymes are agents of almost all chemical transformations in the body.

Chapter Eleven
Microbiology

The term 'microbe' is no longer in professional use but microbiologists (far from being, as their name implies, very small biologists) are students of bacteria, protozoa and viruses, all of which are very small. Sometimes they are collectively called 'micro-organisms', but this term is not appropriate for the virus, whose claims to be considered an organism have already been discussed and dismissed (see pp.8,9); we also have some reservations (below) about the inclusion of protozoa in this category.

However, microbiologists are united not so much by their subject matter as by the nature of the problems which interest them: the structure and assembly of bacterial cell walls and the protein capsules in which the unwelcome information conveyed by viruses is wrapped up. In addition, microbiologists are deeply preoccupied by problems to do with information storage and transfer in micro-organisms, the nature of bacterial adaptation and the change brought about in the cell surface by infection with a virus.

Bacteria are usually capable of surviving outside the body in suitable nutrient media, in which their growth is limited only by the virtual impossibility of eliminating the toxic waste products of their growth as fast as they are formed. Some organisms are exceptional: the causative agent of leprosy, *Mycobacterium leprae*, is an obligative cellular parasite — it will not grow in a cell-free medium, but only in association with living cells. Bacterial organisms have almost prodigious physiological versatility. Bacterial populations can be coaxed to grow under the most unlikely conditions,*

---

* Some can even live in and on gasolene and kerosene.

and those adapted to grow at relatively high temperatures play a specially important part in the manufacture of the so-called 'biological' detergents. Moreover, bacterial populations soon acquire resistance to antibiotics such as penicillin and streptomycin. The term 'bacterial adaptation' used to be described as the 'training' of bacteria both to make use of new foodstuffs and to resist the action of antibiotics, but the word is a very unfortunate one because it suggests an adaptive response on the part of the individual organism. A very distinguished English physical chemist went to some lengths to demonstrate that the 'training' of bacteria was indeed a Lamarckian or instructive process – one in which the new source of nutriment or the new antibiotic evoked from each organism a specific adaptive response that coped with the new situation. But as explained in Chapter 5, it is now beyond reasonable question that these processes are Darwinian and not Lamarckian in character – i.e. genetic variants are already present in the population and in the adaptive process, and their proliferation is so far favoured selectively that in due course they become the predominant type in the population: such phenomena directly rebut the weary old protest that the process of evolution has never been shown to occur in nature. Many bacteria manufacture *antibiotics* – synthetic products that interfere with or altogether impede the growth or multiplication of other bacteria. Some have been turned to very good uses medically (penicillin, streptomycin), but their precise function in the natural history of bacteria is not exactly known: it may however be guessed that in enormously complicated bacterial communities, such as those of the soil and those that live in the rumen of ruminant animals, antibiotics play an important part in the balance of power between one bacterial species and another. Most antibiotics are very toxic to human beings and other mammals because they interfere with growth processes that bacteria share with the ordinary proliferating cells of the body. A few, however, are relatively non-toxic – notably penicillin – because they interfere with the manufacture of substances peculiar to bacteria, so that ordinary body cells are not affected.

The occurrence of muramic acid in organisms such as

spirochetes, formerly classed as protozoans, is now taken as strong evidence that they are better classified as bacteria. Attempts to classify bacteria along the lines of the grand taxonomic hierarchy as it applies to higher organisms have not won sympathy, but a generally useful distinction is usually drawn between bacteria that do (gram-positive) or do not (gram-negative) retain a dye (Gentian violet*) that is used to make them more easily visible under the microscope. The very simple sexual process of bacteria does not introduce any innovation of principle to our understanding of how information is propagated from generation to generation or how it is transcribed from one form of nucleic acid into another and eventually translated into the form of protein. It is a matter of real importance to record, however, that the coding system of the bacterial species *Escherichia coli* is essentially similar to that of higher organisms and that much of our knowledge of processes of transcription and translation (referred to in the chapter on molecular biology) was indeed derived through the study of *E. coli*, particularly by J. Monod, F. Jacob and A. Lwoff. Joshua Lederberg was the first to show that bacteria enjoy a sexual process, i.e. a system of genetic recombination, and beyond this they also possess 'parasexual' mechanisms by which information can be transferred from organism to organism. The most important of these processes historically is the phenomenon of the transformation of pneumococci, which lies at the root of most of the great triumphs of modern molecular biology and therefore deserves to be described in some detail.

Bacteria such as pneumococci are often classified by the chemical nature of their carbohydrate capsules, which have a profound effect on their virulence and behaviour in culture. 'Bacterial transformation' is the name given to the process by which extracts of pneumococci of one capsular type seem somehow to 'infect' other pneumococci with the power to manufacture capsules of their own type. Dr Fred Griffith, a Medical Officer of the Ministry of Health in London, made a particular study of pneumococcal transformation because,

---

* Gentian violet is ('of course', as they say on the B.B.C.) the name given to an ill-defined mixture of the more highly methylated violet pararosaniline dyes.

being an epidemiologist, he was specially interested in problems to do with the ups and downs of bacterial virulence. A new era in biology began when scientists at the Rockefeller Institute for Medical Research in New York showed that the infective agent was none other than deoxyribonucleic acid, DNA, thus making it extremely likely that DNA was the vehicle of genetic information, a conjecture which all subsequent work has confirmed. Some developments of this original work are described in the chapters on molecular biology and biological inheritance (12 and 3).

Unlike most bacteria, *viruses* cannot be grown away from the living cells they infect. The process of viral infection is essentially a subversion of the host cell's synthetic machinery, which has the effect of causing more virus material to be produced. It is incorrect to speak of viruses as if they were tiny micro-organisms that proliferated in cells using the cell sap as a culture medium — much as bacteria may proliferate in the body fluids or in media outside the body. A virus's genetic information is encoded in one or other of the two main forms of nucleic acid, DNA or RNA. The nucleic acid is wrapped up in a coat of protein. Viruses have a crystalline orderliness and much excitement was caused when W.M. Stanley first crystallized a virus, but the clarification this act seemed to bring with it was short-lived, for at that time it was believed that viruses were proteins and the nature and significance of their content of nucleic acid was not yet understood.

Some viruses — notably those responsible for psittacosis and the eye disease trachoma — are intermediate in size between ordinary small viruses and bacteria. Such intermediate viruses might represent simplified bacteria. Some contain muramic acid and respond to antibiotics, but in the main our recovery from virus diseases is the work of our own antibodies (see Chapter 13, Immunology). However, antibodies cannot get at viruses inside cells and it may be that a more important form of immunity is a cell-mediated immunity, directed against the new antigens which the virus causes to be formed at the cell surface. The audacious view has been expressed that much of the damage caused by virus diseases, particularly those affecting the nervous system, is not so much the result of a directly destructive action of viruses on cells as the effect of the inflammation that accompanies the immunological attack against virus-infected cells. This has been formally shown true of the virus disease in mice known as lymphocytic choriomeningitis. The disease has no grave pathological effect even when free virus is recoverable from the bloodstream (viraemia). The trouble begins when an immunity reaction is mounted against the virus-infected cells. Something of the same kind has been thought on less good grounds to be true of poliomyelitis and multiple sclerosis — a disease whose etiology is not fully understood, but in which there is some evidence of a virus-triggered auto-aggressive process. The interpretation and melioration of multiple sclerosis is in our opinion one of the most urgent problems in medical microbiology in the western world, not so much because of the frequency of the disease as because of its life destroying property.

Another large class of disease-producing micro-organisms, which has come into increasing prominence in the past few years, is represented by the *mycoplasmas*, sometimes referred to as PPLOs, standing for 'pleuro pneumonia-like organisms'. Mycoplasmas have the property of appearing where they are unexpected and not wanted, e.g as passengers in tissue cultures thought to be free of all infection. The full extent of the damage they cause is not yet known.

Infection by one virus impedes infection by another. A chemical agent is responsible: Interferon, whose clinical promise as a prophylactic has yet to be fulfilled.

An understanding of the phenomenon of 'competitive inhibition' is one of microbiology's many contributions to general biology. The antibacterial Sulphanilamide, so P. Fildes and D.D. Woods found, acts by mimicking an essential growth factor, *para*aminobenzoic acid, so closely that it is used by many micro-organisms but, not being quite right, it interferes with their proliferation. It is now thought that many antibiotics and antiproliferative agents (see Chapter 14, Cancer) act as competitive inhibitors.

In any literal sense *protozoa* are micro-organisms, but an academic convention has grown up according to which they are studied by persons describing themselves as protozoologists or parasitologists rather than as microbiologists. Protozoa roughly correspond in structure to single cells of higher organisms, though in general they are much larger and have much more complicated cellular organelles.

Among the problems invented by nature-philosophers to give biology an unnecessary air of profundity (it is difficult enough already) is the problem of whether a protozoan is to be regarded as a single cell or as a non-cellular whole organism. Nothing turns on the answer: they can be thought of either way.

The most famous protozoan is that which is known to all laymen as 'the amoeba' — most famous because of the popular belief that it is some kind of primordial living thing that consists of little more than 'a minute speck of protoplasm' (see p.9). It is not even all that minute, for young eyes can see one of the large *amoebae* without means of magnification.

Protozoa are divided into groups roughly according to their method of locomotion. *Amoeba* moves along the substratum with the help of the organs called *pseudopodia*, the prefix *pseudo* presumably serving to correct or rebuke simple-minded folk who might suppose it to have real feet.

Other large groups of protozoa move by means of cilia or flagella respectively. Both words are plural forms, so to refer to 'ciliae' and 'flagellae' is to invite the superior titters that keep tyros in their place. The cilia of protozoa normally cover the whole surface and are thus very much more numerous than flagella, of which there are normally only one or two, arising from the apex of the cell. Both are extensions of the relatively firm outer surface of the cell, and both work in the way described on pp.128–9.

Protozoa are responsible for what is quantitatively the most serious of all infectious diseases — malaria — but if small organisms were to be rated according to their contributions to our understanding of biology they would rank far below bacteria.

Protozoa should not be thought of as modern representatives of our remotest evolutionary ancestors, but rather as successful organisms that have struck out on evolutionary lines of their own and have found special ways of solving the problems that confronted them. Bacteria, however, may in just one respect retain a vestige of the remote past: the idea has been gaining ground that the little organelles in cells known as mitochondria, which are the seat of cellular oxidation, represent the present-day relic of some ancient and evidently fruitful symbiotic association between bacteria and body cells. Mitochondria do indeed possess, unexpectedly, a ration of DNA of their own and their response to inhibitors of protein synthesis is in some ways akin to the response of bacteria.

Viruses are only recognizable when they are cytopathogenic, i.e. cause pathological changes in cells, but there is no reason why all viruses should be so. It is quite possible that our cells harbour a number of silent viruses that have already yielded up their store of genetic information, which has been incorporated into the genome of the cell.

The enforcement of public health measures will in the near future lead to the disappearance of smallpox and poliomyelitis and the loss of the genetic information that specifies their synthesis — except in so far as the viruses may be preserved in a museum built for the purpose. The institution of a museum to preserve vanishing pathogenic viruses may seem to carry sentimentality too far, even in an age notorious for the cultivation of nostalgia, but it is in reality quite a good idea, for if some virulent mutant virus of either kind were to arise there might easily be some useful antigenic overlap (see Chapter 13, Immunology) between the new virus and the old which could be made the basis of a vaccine.

Chapter Twelve

Molecular Biology

With a note on revolutions in biology

Anyone with a more than superficial interest in biology knows that molecular biology has been, and still is, the greatest success story in biology since the formulation of the theory of evolution. The importance of molecular biology and the undisguised hubris of its practitioners has aroused envy and resentment in many old-fashioned biologists, and though it cannot be denied that there is a certain *nouveau riche* air about molecular biology, of the riches there now is no serious question.

Gunther Stent, one of its most discerning historians, distinguishes two main streams of development in molecular biology — one structural and very largely European and the other informational, i.e. having to do with the storage and transfer of information in biological systems, perhaps carried farthest forward in the U.S.A.

Structural molecular biology may be said to have begun with the audacious application by W.T. Astbury of X-ray crystallographic techniques to such biological objects as feathers, hair, tendon and the fibres that make up a blood clot. The great importance of Astbury's work was to reveal that these biological structures have an essentially crystalline orderliness and regularity, a discovery which gives special point to Schrödinger's famous epigram that the 'so-called amorphous solids are either not really amorphous or not really solid'; Astbury's work abolished for ever the idea of a hard and fast line of demarcation between physical objects and the substances of the living world — a conceptual revolution of the same kind but far greater than that said (in retrospect) to have been brought about by Wöhler's reputed synthesis of urea in 1828.

The elucidation of the structure of proteins — work which will one day make it possible to interpret most of the bio-chemical performances of the body in molecular terms — grew out of a number of distinct discoveries: one was the development by Sanger of chemical methods for 'sequencing' proteins, i.e. for spelling out the arrangement along their polypeptide backbones of the various amino acids of which proteins are compounded, and so making it possible to define the structure of proteins in terms of the linear order known as its 'primary' structure. The secondary structure is that which completes the specification of the backbone, especially as it relates to its branching. The tertiary structure of a protein is the pattern of folding, bulging, etc. and the complete speci-fication of a protein as a three-dimensional structure. When the work of Perutz and Kendrew on myoglobin completed our understanding of the structure of a protein molecule for the first time, molecular biologists recognized this as a mile-stone in the history of biology. Since then, R. Valentine and H.G. Pereira have pretty well elucidated the structure of a virus, adenovirus 12.

W.T. Astbury tried his hand at ascertaining the structure of nucleic acids, but X-ray crystallography and the decoding of crystallographs were too rudimentary and the methods of preparation of DNA then in use far too crude for him to have succeeded. Final elucidation of the structure of DNA grew out of chemical investigations by Erwin Chargaff, and the crystallographic analyses of Crick and Watson at Cambridge and Wilkins and Franklin at King's College, London. The elucidation of double-helical structure was a double triumph, because it both solved the crystallographic problem and provided intelligible structural basis for the interpretation of DNA's unique function as a vehicle of genetic information.

The informational stream in the history of molecular biology began with the discovery by Avery, Macleod and McCarty that deoxyribonucleic acid — DNA — fulfilled some of the qualifications of a biological philosopher's stone, for it turned out to be a DNA that was responsible for the transmutation of one pneumococcal variety into another (see Chapter 11, Microbiology) — something that had, until then, been a most perplexing phenomenon. In a few years it became clear from research on bacterial viruses and in orthodox genetics that DNA is indeed the repository of genetical information and the agency by which it is communicated from one generation to the next. Except for the small quantities of DNA in mitochondria the DNA of animal cells, including the germ cells, is confined to the nuclei, as we should expect, having regard to the fact that the nucleus of the germ cells is the only material nexus between one generation and the next.

It is an accepted generalization in molecular biology that genetically coded information can flow only in the direction of nucleic acid → protein and not the other way about — a view long resisted by conventional biochemists, who were most reluctant to see proteins dethroned and denied what seemed to them a natural right to be credited with the execution of all biochemical performances. According to this revolutionary view, nucleic acids must specify the structure of all the proteins made in the cell. A substantial fraction of molecular biology — that which is sometimes called 'molecular biochemistry' (it would be hard to imagine a non-molecular biochemistry) — has been devoted to working out the means by which the structure of a nucleic acid is eventually mapped into the structure of a protein.

Inasmuch as there are more than twenty different amino acids in proteins but only four different kinds of nucleotide in DNA, more than one nucleotide — in fact, a triplet — must code for each amino acid. The mapping process takes place at two stages, each involving the transfer of information to a nucleic acid. These are *transcription*, in which the nucleotide sequence of DNA is mapped uniquely into the nucleotide sequence of *messenger RNA*, and *translation*, mediated through *transfer RNA* molecules that recognize both amino acids and the RNA code and can thus assemble the amino acids in the right linear order. This translation of genetic into structural information is irreversible, so there is no known or at present conceivable method by which germinal DNA could be imprinted with information acquired in an organism's own lifetime.   This is the principal technical reason why no one now believes in the feasibility of evolution in the Lamarckian style (Chapter 5).

Molecular biology has established a hegemony so comprehensive that an eminent biologist has been heard to remark, 'Nowadays not even a museum attendant can hold up his head unless he calls himself a molecular taxidermist.'

A note on revolutionary movements in biology

Most criticisms of molecular biology as it is today come from practitioners of biology who exercised an equally oppressive doctrinal tyranny in their day. In the great post-Darwinian revival of zoology it was believed by nearly all serious biologists that the really important task was to annotate in ever-increasing detail the course of evolution.   In those bad old days quite a number of biological disciplines were thought to achieve respectability for the first time when it became possible to describe them as 'comparative'.   We remember a time when parasitology was ennobled to 'comparative parasitology' because it provided some testimony about the evolutionary process; the same became true of immunology when in the work of Nuttall and Boyden it began to be used as a method of defining evolutionary relationships.

In the reaction against comparative anatomy, *comparative physiology* became an equally great — or even worse — nuisance. Comparative anatomy being now discredited, the foremost obligation of zoologists was thought to be an understanding of the physiology of lower animals. Much of this work was very dull and unilluminating, and the comparative physiology movement was propagated in England by the appointment of comparative physiologists to the headships of most zoological departments — very often departments formerly presided over by comparative anatomists.   All too often piffling physiological experiments were carried out on lower animals in departments which, by their situation and personnel, could have made important contributions to general biology.

Unhappily, it seems to be of the nature of academic revolutions to become an abuse before their pretentions are finally repudiated with a kind of weary disgust.

The same is almost certain to happen with molecular

biology, bright and shining as its future now seems: a time will come when the sequencing of a protein — the more bizarre the better — will be regarded as an intrinsically praiseworthy procedure and one that entitles its practitioners to rapid preferment and the highest academic honours.

One cannot predict what the character of any new revolution will be, but one can hope that there will be a revolution of understanding in some area of biology where understanding is still lacking — especially the area of memory storage and the processes by which the programmatic element in behaviour is propagated from generation to generation.

Chapter Thirteen

Immunology

The explosive growth of immunology during the past ten or fifteen years has been one of the most remarkable phenomena in modern science, although from the standpoint of the history of ideas its rapid growth in recent years is not nearly so surprising as its almost complete intellectual indolence from about 1900 until the mid-1930s. The reason, we believe, is this: until the mid-1930s or thereabouts, immunology was treated as an appendage to bacteriology from the point of view of research, teaching and university administration. Immunology was therefore confined to its immediate practical applications: it was a matter of vaccines, skin tests, diagnostic antisera, blood groups, allergic reactions and not much else. It had indeed every imaginable shortcoming of an 'applied science' pursued without regard to its deep theoretical foundations. A new era began when chemists, zoologists and geneticists started to build up an entirely new conceptual structure for immunology, whose pillars were to be a study of the biology of self-recognition, the molecular basis of specificity and the process of information transfer in biological systems. To these basic researches we may add an examination of the possibilities of immunology for transplantation, the treatment of cancer and a finer genetic fractionation of the population than any other method has made possible.

An immunity reaction is an adaptive response of an organism which has the effect of destroying, neutralizing, sequestrating or annulling the effects of some foreign

---

* In this chapter, 'I' refers to the author (P.B. Medawar) who is still engaged in immunological research.

intruder such as a bacterium, virus, protozoan parasite, a graft from another individual and quite possibly a malignant growth (see Chapter 14, Cancer). Substances that arouse an immune response are described as antigenic or substantively as 'antigens'. The distinguishing characteristic of an antigen is its foreignness, i.e. its property of being non-Self. However, even Self-constituents can sometimes arouse an immunity reaction, for the foreignness that is a qualifying property for being an antigen means only foreignness to the reacting system, and if some parts of the body have been sequestered throughout life, their liberation through injury or some degenerative process enables them to exercise their antigenic power. Reactions upon such Self-constituents are referred to as 'autoimmune'. Bodily constituents altered by chemical action or virus infection may also arouse autoimmunity.

The pattern of immunological reaction which has come to be regarded, almost certainly wrongly, as typical — indeed as prototypical — is that in which an antigen excites the formation of an 'antibody'.* An antibody is a protein circulating in the bloodstream, whose structure is exactly complementary to that of the antigen or, more accurately, to that part of the antigen which confers its antigenicity upon it. When antigen and antibody meet, any one of a whole variety of reactions may take place: agglutination or clumping if the antigens are in the form of cells, sometimes a rupture of the cell membranes, a co-precipitation when the antigen is itself a soluble protein, or a detoxication when the antigen is a poisonous substance — a toxin — produced by a bacterium. The end result, in a successful immunological reaction, is a destruction of the antigen or an annulment of its effects. The teleological rubric 'foreign = bad' has been pretty dependable, for immunological activity is now known to be essential for life: some children are born without the capacity to manufacture antibodies and they can be kept alive only by antibiotics and massive transfusions of normal blood or of its antibody-containing constituents.

---

* The terminology of immunological reactions is simplistic to the point of idiocy. Medical students are well known to feel embarrassed when they first come to learn it.

Two important manifestations of immunological reactions mediated through antibodies are *lysis*, the physical rupture of antigenic cells and the liberation of their contents into the surrounding medium, and *phagocytosis*, a process in which antigenic particles are engulfed into and very often digested by, or in any case rendered harmless by, 'macrophages' and 'polymorphs' (see p.136, *Blood cells*). The process of lysis is executed by a very complex and unstable blood constituent known as *complement*, which will not work unless it is activated by a specific antibody; and the process of phagocytosis is greatly facilitated and speeded up when the particles which are to be engulfed into the phagocytic cells are coated by a specific antibody directed against them. 'Stimulate the phagocytes' — the sovereign remedy propounded by Sir Colenso Ridgeon in Shaw's *Doctor's Dilemma* — is thus for the most part a matter of stimulating the formation of specific antibodies.

Cellular immunity. For very many years antigens and antibodies dominated the thoughts of immunologists so completely as to exclude any other conception of how the immunological response might work, but nowadays it is realized that an entirely different kind of immunological reaction is mediated through the action not of antibodies, but of lymphocytes (see below) which are hostile to and bring about the destruction of the antigen or whatever its vehicle may be. The reactions mediated through the action of lymphocytes are just as specific as those of which the agents are circulating antibodies. Cell-mediated immunity is that which is responsible for graft rejection (see below), for the destruction of intracellular parasites, for antitubercular immunity and possibly also for antitumour immunity (see Chapter 14, Cancer); on the other hand humoral — that is antibody-mediated — immunity is thought to be responsible for coping with most ordinary bacterial and viral infections.

Lymphoid system. Cells responsible for recognizing antigen and for reacting upon it in one way or another belong to the lymphoid system, a system of cells of which the prototypical member is the *lymphocyte*.

Recognition of antigen. Even with antibody formation — basically a much simpler process than cell-mediated immunity

— the informational problem of how an antigen is recognized and of how a structure exactly complementary to it is then synthesized is one of which the solution was very far from obvious. The problem would not be so difficult if antigens were limited in variety and the antibody formed was one or other of a limited repertoire, but the trouble is that antigens are as various as living organisms themselves, and an organism can manufacture antibodies against antigens which have not even been invented yet — e.g. the synthetic organic chemicals which are being produced yearly in ever greater profusion and which are perfectly capable of conferring antigenicity upon any macromolecule they attach to. Before the coming of molecular biology the problem did not seem to be a dismayingly complex one, for it was believed that an antigen could itself inform the structure of an antibody, i.e. it could itself provide the information which directed the synthesis of an antibody molecule, and thus the antibody molecule could be built around and upon the antigen in such a way as to have an exactly complementary structure. Linus Pauling, the great theoretical chemist, at one time believed that such a process was possible and he devised an ingenious model by which antibody molecules would reshape themselves in the presence of an antigen. This represents the commonsense, instructive or 'Lamarckian' interpretation of the process, but when it became clear that information cannot flow from protein to protein, but only from nucleic acid to proteins, some other explanation had to be sought. One possibility is that lymphoid cells in the course of their development become randomly diversified so that to every possible antigen there corresponds a lymphoid cell, and therefore a potential 'clone' or cell division lineage of lymphoid cells capable of manufacturing the antibody complementary to it. This diversity could arise mutationally in the line of cells that leads from the fertilized egg to the billions of lymphocytes of the adult body, or the mutational process could be one that takes place in some lesser cellular component than the nucleus. At any event, the process envisaged is one in which lymphoid cells with a vast number of reactive potentialities are already available and waiting only to be aroused into full activity by confrontation with an antigen. This is evidently a

Darwinian explanation — i.e. one based upon random variation and selection — and it is significant that one of the very first scientific papers on the subject, by Nils K. Jerne, was entitled 'The natural selection theory of antibody formation'. It is for this reason that the profane immunologists referred to earlier (p.47) used the term G O D as a convenient acronym for 'Generator of Diversity'. The occurrence of random variation in the line of cells arising from the zygote to bring forth adult lymphoid cells is not the only possibility, however; it is at least conceivable that all the information necessary to underwrite the formation of antibodies is already present in the fertilized egg and is therefore equally apportioned to all the lymphoid cells that descend from it. At first this sounds rather like a tall order, but it ceases to be so when one reflects upon the enormous variety and subtlety of information already coded within the zygote nucleus — e.g. upon the fact that the complete repertoire of female behaviour in so far as it is programmed is encoded in the male zygote and *vice versa*, waiting to be released or realized by the appropriate external stimuli, or upon those almost incredibly delicate nuances of the developmental process which make it possible to say that an infant has got exactly his mother's nose or his uncle's habitual leer. If this interpretation is true it follows that when an antigen acts upon lymphoid cells it acts in a manner pretty closely analogous to that of an embryonic 'inducer' (see Chapter 9, Development), and like an embryonic inducer it arouses one rather than another potentiality of the responding cell.

The immunology that grew up in the clear awareness of the famous dogma of molecular biology that relates to information flow was described by Sir Macfarlane Burnet, one of its founders, as 'the new immunology'. In view of the fact that the problems that confront the two sciences are in some ways cognate, it is hardly surprising that microbiologists such as Jacques Monod and Joshua Lederberg were among those who also played an important part in founding the new science. Perhaps Burnet's greatest contribution to biology is that he made immunologists rethink the mechanism of the immunological response in terms of the population dynamics of lymphoid cells and to abandon for ever a

Lamarckian interpretation of the immunological response.

Although the matter is not yet finally settled, it can be taken as certain that *unless* a mechanism is stumbled upon that no one has yet even dreamed of, an immunological response to no matter how strangely exotic an antigen represents an awakening or activation of some pre-existing potentiality in the responding cell and is not in any sense an indoctrination of the cell by some molecular property of the antigen.

Transplantation immunity and tolerance. The chemical make-up of one human being's organs must be so closely similar to another's that the phenomena of transplant rejection may seem specially surprising — so surprising, indeed, that for very many years surgeons did not believe it. They believed, as some surgeons in Russia still believe, that the failure of a transplant was to be attributed more often to faulty surgery than to the extreme precision of the body's ability to identify 'non-Self'. To say this is not by any means to depreciate the importance — as it has turned out, the increasing importance — of surgical expertise in the execution of transplants. Nevertheless, the immunological element outweighs all others in importance. Other factors that play a very important practical part are whether or not the operation is physiologically feasible and whether or not an organ-substitute can be rigged up to tide a patient over what otherwise might be a period of acute deprivation in which his condition would deteriorate seriously. The transplantation of brains belongs strictly to science fiction: it is not possible today, and there is no serious possibility of its becoming possible in the future. As to organ substitutes: heart transplantation would not of course have been possible unless apparatus was available to maintain the process of respiration. There is unhappily no substitute for the liver, although attempts have been made to use a pig's liver outside the body in much the same way as a patient might use a kidney machine. Kidney transplantation was given a great boost by the invention of the kidney machine because it became then, for the first time, possible to bring a patient with diseased kidneys to the operating table in a condition fit enough to withstand the therapeutic procedures to be used on him. Among these therapeutic procedures — and that upon which the whole of modern transplantation rests — is the use of *immunosuppressive agents*, drugs that depress the immunological response long enough to tide the patient over the period during which rejection might occur and at the same time to help the graft to start to undergo the rather mysterious adaptative process which ends by making it fairly acceptable to the body. Most of the immunosuppressive agents used in transplant surgery are agents that prevent cell division, and thus prevent the full expression of the immunological response. With very few exceptions, they were plundered from the huge armoury of cancer chemotherapy. The drug now in most general use throughout the world was introduced by Professor R.Y. Calne after lengthy trials of kidney transplantation in dogs. This is the drug Imuran, a derivative of the drug known as 6-mercaptopurine, which mimics one of the essential ingredients of nucleic acid closely enough to derange cell division. In addition to Imuran, it is customary to use steroid drugs related in their mode of action to cortisone; the dosage of steroid drugs is always raised if there is any suspicion of a transplant's being rejected, because steroid drugs are able to reverse the rejection process. The use of steroids is, however, fundamentally objectionable because of their highly unpleasant side-effects, and the whole immunosuppressive procedure still leaves very much to be desired. The entire dosage regimen is a knife edge in which underdosage will lead to the rejection of the graft and overdosage to secondary damage on cells other than those that transact immunological responses — particularly blood-forming cells; in addition to this there is the ever present danger that immunosuppression may so far weaken the body's immunological defences as to make the patient an easy prey to infectious illness. The trouble is that with one exception no real immunosuppressive agent is known, i.e. no agent that suppresses an immunological response without suppressing a great many other things as well. The one exception is *antilymphocyte serum*, which effectively eliminates the kind of immunity described above as cell-mediated — that which causes transplant rejection. Unfortunately, antilymphocyte serum has grave shortcomings of its own, not the least of which is that if it is to be supplied at an economic price a course of injections costs in the neighbourhood of £1,000. Even in countries without a national health service, means would doubtless be found to get round this difficulty if it could be shown that antilymphocyte serum or the active protein extracted from it was unconditionally safe and very much more effective than any combination of agents already known.

Techniques of immunosuppression are constantly being improved, and it may be taken for granted that if there is a demand for them, heart, liver and lung transplantations will become as frequent and as successful as kidney transplantation is today. Oddly enough, the one tissue which has so far resisted all attempts to transplant it from one person to another is skin. It is either particularly adept at arousing an immunological response or particularly vulnerable to it.

Another scientific advance which will make an important contribution to the success of transplants is the development of tissue typing, i.e. the grouping of individuals for immunity provoking factors in a way analogous to the grouping of blood for transfusion. Research in this area is led by Ceppellini in Turin, Dausset in Paris, Amos at Duke University, van Rood in Belgium, Batchelor in England and one or two others. From time to time all workers on tissue grouping convene to exchange research results and opinions — sometimes, understandably enough, the opinion that their own system of nomenclature is very much superior to anyone else's. Tissue groups differentiate the members of a human population even more finely than blood group differences, and give rise to a genetic variegation more close-textured and fine-grained than any other. For this reason, grafts between two different human beings invariably fail unless immunosuppressive agents are used or the donor happens to be the recipient's identical twin. For purely genetic reasons, grafts transplanted between relatives usually do better than grafts from unrelated donors.

Donor supply is a continuing headache in kidney transplantation and is unfortunately worse with heart and liver transplantation. It would be greatly relieved if only methods

could be devised of storing tissues without deterioration at
−70° or −190° Centigrade − temperatures chosen, not
arbitrarily, but because they are the easiest low temperatures
to achieve as a matter of routine (being those of solid carbon
dioxide and of liquid nitrogen respectively).

Blood groups are a form of inborn differentiation between
human beings similar to that which normally prohibits the
transplantation of organs from one human being to another.
The differentiation of humans into groups A, B, AB and O
is familiar to almost everybody by name because of their
well-known importance in blood transfusion. They were,
moreover, the first blood groups to be discovered, but since
then systematic research has so greatly enlarged the known
variety that by adding such groups as Kell, Duffy, MN and
all the many variants of the Rhesus groupings, we can say
with some confidence that the number of combinational
differences between known blood groups exceeds the number
of people alive in the world today. The letters A and B
stand for antigens present on red blood corpuscles. These
antigens do not normally have an opportunity to arouse
immunity *de novo*; on the contrary, the antibodies that
correspond to them are found readymade in the individuals
that lack them: a person of blood group B has the antibody
anti-A circulating in his blood serum and so conversely for a
person of blood group A. Someone of blood group O (lacking both antigens) has antibodies directed against both A and
B cells. When suspensions of red blood corpuscles are mixed
with the antibodies directed specifically against them, they
are agglutinated − clumped together. This is a very serious
impediment to blood transfusion, for grave accidents would
occur if for example someone of blood group B were to
receive a transfusion of group A blood and *vice versa*; on the
other hand, an individual of group AB can receive blood from
donors of either kinds and a member of group O is a universal
donor. Antigens of the rhesus series* are notorious for their
ability to cause haemolytic disease in the newborn, which can

---

* The odd name rhesus derives from the fact that the rhesus blood groups were
discovered by immunizing rabbits with the red blood corpuscles of a rhesus
monkey, so producing an antiserum which reacted upon approximately 85
per cent of human bloods and had no effect upon the remaining 15 per cent.

afflict children of higher birth rank in a minority of marriages between rhesus negative women and rhesus positive husbands. What seems to happen is that at some stage of pregnancy, most commonly perhaps at parturition itself, fetal blood corpuscles containing one of the stronger rhesus antigens enter the blood circulation of the mother and thus immunize her. The mother is now immunologically forearmed by much the same kind of mechanism as that which normally prevents a patient getting measles twice, in the sense that any exposure to rhesus antigens such as may occur in a later pregnancy produces a stronger and quicker antibody response. The antibodies so formed may enter the circulation of the unborn child and inflict great damage on its red blood corpuscles, a process producing jaundice and at the same time stimulating the proliferation of the cells which manufacture red blood corpuscles — 'erythroblasts' — so that the term 'erythroblastosis fetalis' is commonly used to describe haemolytic disease caused by rhesus incompatibility. Sensitization by rhesus antigens of a mother known to be at risk can be prevented by an ingenious procedure devised by immunogeneticists: the deliberate injection into the mother of anti-rhesus antibody which will promptly destroy the red blood corpuscles carrying the rhesus antigen when they enter the maternal circulation.

Teleology of the blood groups. The raison d'être or survival value of differentiation into blood groups, that is the special function blood group differentiation fulfils, is not yet fully understood and no one quite knows in what way the sub-division of human beings into different blood groups is of any use. It is at least relevant, however, that sensitization of a rhesus negative woman by a rhesus positive fetus is less frequent when mother and fetus are incompatible with respect to the ABO blood groups. The reason for this may be that if, for example, in addition to being rhesus positive the red blood cells of the fetus are of blood group A and the mother is of blood group O or B, then the red blood cells escaping into the circulation will be promptly destroyed by the mother's readymade anti-B antibody. Thus the differentiation into blood groups of one kind may annul the harmful effects of differentiation of another kind. This does not, however, provide a general explanation of blood group diversity.

The notion of tolerance. It was brilliantly perceptive of the microbiologist Macfarlane Burnet to realize that in the development of the cells that form the immunological response system, some mechanism must exist by which any tendency on the part of the system to react upon Self-constituents is suppressed. Burnet's idea was that if an antigenic substance were presented early enough in life to an embryo, it would be accepted as a Self-substance and not reacted upon later in life. In forming this opinion he was strongly influenced by the work of the distinguished agricultural geneticist Dr Ray D. Owen of Madison, Wisconsin, on the extraordinary properties of cattle twins. Cattle twins, like human twins, are of two kinds: on the one hand identical twins which are genetically identical and necessarily of the same sex, having started life as a single individual and through some otherwise innocuous developmental accident become two, called *monozygotic* or *one-egg* twins, and on the other hand *dizygotic* — *two-egg* — twins which each arise from separate fertilized eggs and which are no more alike than ordinary brothers and sisters — they are litter mates in a litter of two. Owen's extraordinary discovery was that each twin contained red blood corpuscles of two distinct antigenic types: one type its own and the other type its partner's. The technical name for an animal containing cells descended from two distinct fertilized eggs is a *chimera*. Dizygotic cattle twins are chimeras because of the intermixture of each other's red blood corpuscles. Owen clearly saw that this extraordinary state of affairs was made possible by the fact that cattle fetuses share a placenta, with the effect that the two blood systems communicate with each other and a free exchange of blood between the twins is possible. This then was a natural experiment which might have been designed to test the validity of Burnet's hypothesis: each twin had confronted its partner with its own distinctive antigens very early in embryonic life and had therefore, according to Burnet's hypothesis, switched off its twin's capacity to react upon the other's cells. For this reason the blood cells and blood-forming cells exchanged in fetal life could survive up to and far beyond birth and so give rise to and maintain the chimeric state. The interpretation was made still more convincing when Rupert Billingham and I and two young colleagues from the Agricultural Research Council showed that dizygotic cattle twins would accept grafts of each other's skin, although cattle in general react rather violently against skin grafts transplanted from other members of their own species. This mutual tolerance was highly specific, for the twin that would accept its partner's skin graft would throw off in ten days a skin graft transplanted from a third party. In due course workers in Copenhagen showed that dizygotic cattle twins would accept each other's kidneys too, so evidently the mutual transfusion of blood before birth had switched off each twin's power to react upon tissues other than blood.

Human twins are sometimes, very rarely, chimeras and they too, as Woodruff showed, may accept grafts of each other's skin. So are, and so do, chicken twins — chicks that hatch from double-yolked eggs. Thus the phenomenon of mutual tolerance in chimeras is not an idiosyncrasy of cattle.

Nevertheless, Burnet's notion was not universally accepted until it became possible to reproduce by deliberate experimentation the phenomenon that occurs through a natural accident in twin cattle. What would an adult animal's reactivity be, it was asked, towards an antigen of which it had had experience very early in embryonic life? Several groups of workers tried to find out and, as luck had it, Burnet's own experiments failed because the antigens he used in order to induce a state of non-reactivity in later life could not persist in the organism for the time now known to be necessary to produce and maintain the tolerant state. No such difficulty faced those of us who worked with living cells, because the cells remain alive, enjoying the privilege conferred by any state of tolerance they might induce. Milan Hašek, a brilliant Czech experimental biologist, succeeded by deliberately twinning chick embryos: using two embryonated hen's eggs he laid bare small areas of the vascular membranes by which the embryo respires and made a vascular bridge between the two embryos by interposing a fragment of living tissue which was penetrated by blood vessels from both sides. When the chicks grew up they accepted grafts from each other and were quite incapable of reacting upon each other's red blood corpuscles — something which otherwise they would almost certainly have been able to do. Billingham, Brent and I used mice for our experiments, inoculating the embryos of one strain with a potpourri of living cells from mice belonging to another strain, with the effect that when the embryos which had been inoculated grew up they were able to accept grafts from mice of the strain which had provided the fetal inoculum in the first place. We described the phenomenon as 'actively acquired tolerance' on a model of 'actively acquired immunity'.

The methods used to produce transplantation tolerance in experimental animals are clearly not applicable to human beings, and there is no direct relationship of cause and effect between the discovery of tolerance and latter-day triumphs of tissue transplantation. On the contrary, the effect of the discovery was almost wholly a moral one: the experimental induction of tolerance showed — something which until then had been very much an open matter — that the barrier which normally prevents the transplantation of tissues between different individuals can indeed be overcome. Until the discovery of tolerance it would have been perfectly possible to maintain, as some people did, that the ambition embodied in experimental tissue transplantation could never be achieved. The search for ways to get round the transplantation barrier could therefore be loftily dismissed as the pursuit of a chimera — and so it proved to be, for the reasons explained above.

Miscarriages of the immunological process.   Although immunological reactivity is essential for life, immunological processes make themselves more obviously apparent by the way they go wrong than by working correctly. It is not of course fair to include blood transfusion accidents and graft rejections under the heading of maladaptations of the immunological process, because it is our fault when they occur, and they occur only in a situation of our own making. Only in haemolytic disease of the newborn can nature itself be justly incriminated. The main aberrations or maladaptations of the immunological process are allergies, hypersensitivity and autoimmune phenomena. In one form or another all of them can be regarded as the price we must pay for possessing a response system attuned with the exquisite sensitivity necessary to discern and react upon non-Self components. The most familiar allergies are those aroused by pollen grains, danders, house dust and certain dietary constituents: the allergenic capacities of eggs, shellfish, fish and certain fruit are very well known. There is a very strong element of idiosyncrasy in allergic sensitization. Individuals who contract one kind of allergy are more likely than members of the general population to contract another. 'Delayed type hypersensitivities' are allergies of an entirely different class. In their most familiar forms – among which are sensitization to certain industrial chemicals like dinitro-chlorobenzene – they make themselves apparent as raised, hardened, red and sometimes extremely painful swellings of the skin. Some bacterial antigens are notorious for the ability to arouse delayed type reactions and the area of painful redness around a boil very often represents a reaction of this type. The reaction is the cell-mediated type (see p.100). The ingredients of the tubercle microbacterium are specially well able to arouse delayed type hypersensitivity, and advantage is taken of this in the so-called 'Mantoux test' in which a minute quantity of tubercle bacterial extract is injected into the skin: a positive reactor is one who has at some time in the past been exposed to tubercle bacteria and the reaction need not be construed as evidence of active tuberculosis. The study of the tuberculin reaction and of immunity to tuberculosis generally has had a profoundly important influence on the history of immunology: the whole phenomenon is so important and such a puzzle that it attracted many of the very brightest medical students to the study of immunology. More recently the study of transplantation and of tumour immunity (q.v.) has had a rather similar effect. It is not the promise of easy victories but rather the existence of an intellectual challenge that is the most effective form of recruitment into research. Although it sounds sententious, this generalization can be illustrated over and over again in the history of biology: one thinks for example of the recruitment of biologists into embryological research in the 1930s, into ethology in the 'forties and into molecular biology today.

Autoimmune diseases arise, it is thought, from two different causes. The first is a failure, or accidental circumvention, of the natural mechanism of tolerance because a potentially antigenic tissue ingredient, which has hitherto been anatomically or otherwise sheltered from the immunological response system, gets access to it and arouses a response which may in itself cause more damage and thus intensify the response — 'positive feedback'. It is probable that something of the kind occurs in autoimmune thyroiditis (Hashimoto's disease), and almost certain that it accounts for the distressing condition known as sympathetic ophthalmia in which the remaining good eye goes blind some time after a penetrating injury to the other.

Alternatively, autoimmunity may arise through the attachment to normal body constituents of substances or agents that confer antigenicity upon them. This is essentially what happens if the skin is contaminated by the allergenic chemicals mentioned in an earlier paragraph as possible causes of delayed-type hypersensitivity reactions; it is particularly instructive because the chemicals in themselves are often not antigenic, but they become so when attached to bodily constituents. A matter of much greater moment is the antigenicity that body cells — notably nerve cells, neurones — may acquire as a result of infection by a virus. It is quite widely believed that an important fraction of the neurological damage that occurs in poliomyelitis and multiple sclerosis may be the consequence of an autoimmune reaction of just this kind. Other diseases in which autoimmunity is thought to play an important part include rheumatoid arthritis and scleroderma.

The treatment of autoimmune diseases sounds as if it ought to be a straightforward matter of the administration of immunosuppressive agents. However, in reality it is a much more difficult business, for the suppression of antibody formation might easily suppress a mechanism which helps to hold cell-mediated immunity under control, and before undertaking a wholesale suppression of cell-mediated immunity we may well have gloomy forebodings about what the consequences of such a procedure might be. It is for such reasons that institutes exist for the prosecution of 'basic' medical research, the goal of which is not much more revolutionary than to give the physician some rational understanding of what he is doing and how he may most easily achieve the effects he hopes to secure. The barrenness of immunology during the first thirty years or so of the twentieth century, and its remarkable efflorescence since then, can be taken as an object lesson and warning for all those who confidently declare that laboratory science should be confined to the attempt to achieve immediate practical objectives. By the ordinary — though not yet fully understood — processes of scientific research the study of immunology has greatly enlarged our understanding of protein synthesis, the molecular basis of specificity and the means by which information is transferred from molecule to molecule in the body. In addition, it may be credited with the mastery of safe blood transfusion techniques, the transplantation of tissues and the control of at least some of the self-destructive manifestations of immunological responses.

Chapter Fourteen

Cancer

A long tradition of inexplicable origin has it that cancer is a subject upon which biologists have profound and medically important things to say. One can only suppose that it grew out of the illusion that cancers are merely cells which have escaped from the overall growth-controlling influences of the body. To this declaration, a pathologist would be entitled to retort that cancers are not 'merely' anything and that rapid growth is not their most distinctive characteristic, nor that from which most is to be feared. Some of the other characteristics that distinguish malignant tumours are (a) their invasiveness, i.e. their power to infiltrate into the surrounding tissues, escape into the *lymphatics* that drain the tissues and propagate themselves around the body where secondary centres of growth may be set up and (b) possibly their power to circumvent the natural immunological defences of the body (see below).

Cancer is a general name for a great variety of different growths, so that to speak of 'the cause of cancer', or indeed to speak of 'cancer' as if it were a single disease, makes no more sense than to lump together tuberculosis, pneumonia and plague under the heading of 'bacteritis' and to enquire plaintively what is the cause and cure of bacteritis.

Kinds of cancer. Cancers of epithelial tissues, i.e. of tissues that bound surfaces, particularly of the skin and mucous membranes, are known as carcinomas (or by purists — others would say pedants — as carcinomata). Cancers of connective tissue cells or of cells that are normally disperse in character are called sarcomas or — with the same proviso — sarcomata. Tumours of the white blood corpuscles are known comprehensively as leukaemias. A further differentiation is marked by specific mention of the tissues from which tumours arise; thus a malignant tumour of bone is an osteosarcoma and of lymphoid cells is a lymphosarcoma. Tumours of the pigmentary cells of the skin, known as melanocytes, are melanosarcomata and tissues of the supporting cells of the central nervous system are called gliomas. Very often the plain suffix 'oma' is used to distinguish a benign as opposed to a malignant growth: thus an osteoma may be no more than a bony excrescence or protuberance and neuroma no more than a swelling of a nerve caused, perhaps, by some miscarriage of the regenerative process.

Causes of cancer. Transformation of normal into malignant cells can be brought about by very many different agencies. Among them are X-rays or other ionizing radiations and a whole variety of hydrocarbons, many of which were first extracted from or identified in distillates of coal tar, among them methylcolanthrene, benzpyrene and dibenzanthracene. Viruses are well known to cause certain tumours in chickens (a discovery of Peyton Rous) and also to cause leukaemias in mice and cats and probably — though this has not yet been the subject of a completely convincing demonstration — in man. One of the most interesting, though in some ways one of the most bewildering, of methods for producing tumours in experimental animals is insertion into the subcutaneous tissues of a small, relatively impermeable, plastic film. If the film is perforated no tumour forms, so evidently it is not the material of the film itself that can be held responsible: some impairment of cellular or fluid movement is more likely to blame.

Treatment of cancer. Tumours are treated by surgical excision whenever possible — a procedure in which early recognition and the ability to localize the tumour growth by X-radiography are both of the utmost importance. With some tumours extirpation is of course not possible. Among these are leukaemias and tumours so far advanced that secondary growths (metastases) have already established themselves throughout the body. Alternatively, or in addition, tumours are treated by one or other of the great variety of antiproliferative agents — agents that impede cell division (not in malignant tissues only, unfortunately, but in dividing cells throughout the body). X-radiation or gamma-radiation — such as is emitted from a slug of radioactive cobalt — are the most important antiproliferative procedures. Their great advantage is that radiologists have learnt to control the administered dosage very exactly and that, unlike the chemical antiproliferative agents, they can be switched off when they are thought to have done their duty. Chemical antiproliferative agents improve year by year, or almost month by month. Among them are substances that act by mimicking and so impeding the proper use of essential food-stuffs (e.g. the antifolic methotrexate), substances that mimic the structure of essential constituents of nucleic acid, the multiplication of which they accordingly interfere with (substances of this latter category include captopurine and azathioprine).   Substances of both classes illustrate the principle of competitive inhibition (see Chapter 11, Micro-biology).

Another line of attack is through hormones, for some tumours are made up of cells which are hormone-dependent: tumours of this kind can often be kept under control by withdrawing the hormone they need and sometimes substituting for it a hormone with the opposite effect. A case in point is the treatment of tumours of the prostate gland, an accessory male sex organ, by the administration of, for example, the synthetic oestrogenic hormone, oestradiol.

The possibilities of an entirely different style of treatment have been explored rather excitedly in recent years: this is *immunotherapy*, based on our knowledge of immunity to tumours.

One of the earliest discoveries made in the experimental analysis of tumours was that by repeated coaxing (in effect by repeated trials of which many would be failures) tumours arising in rats and mice could often be transplanted to other rats or other mice respectively. It was not very often that the little tumour grafts 'took' and started growing right away; on the contrary the more usual thing was for the tumour first to grow and then to dwindle away·to the accompaniment of a rather fierce attack by the recipient's lymphocytes, cells which play a crucially important part in the immunological defences (see Chapter 13, Immunology). A mouse or rat in which a tumour had first grown and then dwindled away was absolutely refractory to the inoculation of that same tumour on a second occasion. The same refractory state could sometimes be set up merely by the inoculation into the intended recipients of the tumours of a hotch potch of normal tissue cells, especially embryonic cells. Phenomena such as these encouraged early students of tumour transplantation to believe that the regression of a tumour was, in effect, a cure and that forearming of its intended recipient by inoculations of tumour cells and specially of normal embryonic cells was a preventive measure. Peyton Rous (1879–1970), the greatest experimental pathologist of his day, was perhaps the first man to shatter these illusions. He asked what evidence there was that the 'immunity' so commonly spoken of was an immunity directed against the tumour as such, or whether it might not merely be an immunity directed against the tumour graft considered as a genetically foreign cell — whether, in fact, it might not be an immunity similar to that which is directed against foreign grafts generally (see Chapter 13, Immunology).

Peyton Rous's forebodings proved very well founded and no wonder, for in the early days of tumour transplantation inbred strains of mice (p.34) had not yet been developed and the mice actually used in these experiments — 'the grey mouse' and 'the white mouse' — were a mixed bag whose only uniformity — that of colour — was quite trivial. Worse still, the early tumour transplanters were unhappily distracted by all kinds of irrelevancies of no real biological significance, among them allegedly seasonal variations in the growth of tumours. Although they did not realize it, the early students of tumour transplantations were studying not tumours but transplantation — a subject to which they did in fact make a number of very important contributions, particularly in the area of the genetics of transplantation.

A new era began in the 1930s when a number of American cancer research workers discovered true tumour immunity, i.e. an immunity directed against malignant cells as such. They found, in effect, that a mouse sometimes develops immunity even against an autochthonous tumour, i.e. a tumour arising and growing in itself. No such discovery would have been possible before the development of strictly inbred strains of mice — mice which, differences of sex apart, resemble each other as closely as if they were identical twins. Great hopes have been built on this phenomenon of auto-chthonous tumour immunity, and some of these hopes are just beginning to be fulfilled. Tumour immunity is of the same general kind as transplantation immunity, i.e. the process which leads to the rejection of foreign grafts (see Chapter 13, Immunology). However, tumour antigens are not all that easy to identify and define by orthodox immuno-logical methods and it is a cloud on the horizon that there has not yet been a finally conclusive demonstration of an immunological response to human tumours generally. Circumstantial evidence in favour of human tumour immun-ity is nevertheless strong enough to have persuaded most pathologists to have accepted the idea on probation, and it is therefore worth examining some of its practical consequ-ences. The most important are:

1. The regression of tumours, so far from being a great rarity and something only reluctantly believed in, must be quite a common phenomenon. Indeed, many more tumours must arise than we ever become aware of in a clinical sense — tumours which have been spied out by circulating lymphocytes and which have aroused and succumbed to an immunity reaction.

2. It will be worthwhile to remove the greater part of a tumour or as much tumour tissue as is accessible even in the clear realization that the whole of it cannot be remov-ed. The rationale of doing so is to remove the burden thrown upon the subject's immunological defences by the constant outpouring of antigenic matter from the tumour — a process which may very well impede the immune reaction.

3. The centre of gravity of clinical cancer research will change from the empirical trial of antiproliferative agents to an enquiry into why immunological processes that ought to work very often don't and into how the immune process can be boosted.

4. Cancer chemotherapeutic agents which are also

immunosuppressive agents (and most of them are) should be used with the utmost circumspection; moreover, it should not be taken as a matter of course that in the removal of a regional tumour all the regional lymph nodes (see Chapter 16, Circulatory Systems) should be radically extirpated. Indeed, in the extirpation of mammary tumours there is no convincing evidence that wholesale removal of all draining lymph nodes is a beneficial procedure.

Early diagnosis of cancer: fetal antigens. The entire clinical management of cancer would be transformed if only it were possible to secure early evidence of the upgrowth of a malignant tumour in the body — particularly a tumour of internal organs. Quite a high proportion of cancer research is directed towards a solution of this problem. One lead that has been followed up turns on the identification in the body fluids of minute quantities of fetal or embryonic substances.

The rationale of this otherwise strange-sounding approach is that fetal substances which make a transient appearance in development are sometimes manufactured anew in malignant cells — as if it were a regular accompaniment of the malignant transformation that genes which operate in fetal life and are then normally switched off are somehow reawakened or 'derepressed' in tumours. The occurrence of this phenomenon is not to be taken as evidence of the old-fashioned 'embryonic rest' theory of tumours, which supposes that tumours arise from embryonic cells which have somehow got left out in the course of developmental processes, only to come to life at a later stage and grow vigorously. This theory has not stood up to critical tests and is no longer seriously entertained.

Immunological surveillance. The teleology of graft rejection has always been obscure, but most people are content to regard it as the price paid for possessing an immunological system with exquisitely discriminating powers of distinguishing Self from non-Self.

An entirely new way of looking at the problem has grown out of the speculations of Lewis Thomas and Macfarlane Burnet: the rejection of grafts is a tiresome byproduct of the existence in the body of a monitoring system exquisitely tuned to prying out and eradicating abnormal variants among the body cells. Needless to say, the immunological theory of defence against tumours has given the notion of immunological surveillance extra credibility.

Unhappily, all is not plain sailing: the mutant mice known as nudes in which cell-mediated immunity is so gravely impaired that even grafts of human tissues can be accepted are not nearly as susceptible to tumours as the theory of immunological surveillance encourages us to believe. The same goes for mice artifically deprived of the thymus in infancy and thus also barely capable of mounting a cell-mediated immune response. However, the idea that there is something in tumour immunity and probably in the notion of surveillance as well is supported by the fact that mice in which immunological capability has been raised well above the normal level (as shown by independent tests) are much more resistant than normal mice to the growth of autochthonous tumours.

The upsurge of new ideas in the area of tumour immunology is the standing reproach to all counsellors who have maintained that what is needed in cancer research is not so much money as ideas. Modern cancer research, like immunology itself, attracts very many brilliant young scientists, full of ideas, but mainly through lack of funds not all of these can be tried out. However, it should be clearly understood that *the* cure of cancer is never going to be found. It is far more likely that each tumour in each patient is going to present a unique research problem for which laboratory workers and clinicians between them will have to work out a unique solution.

Chapter Fifteen

Cells and Tissues

A cell may be thought of in two rather different ways. On the one hand, it may be seen as the smallest sub-division of the body that is capable of independent life, although 'independent' life is a rather far-fetched description of the very nicely adjusted surroundings which must be provided for the cell if it is to remain alive after removal from the body. On the other hand, the cell may be regarded as the parish or as the administrative domain of a nucleus. When a spermatazoon and an egg cell unite to form the fertilized egg, the *zygote*, from which sexually-reproducing organisms develop, their respective nuclei fuse into one and combine their genetic information. In the earliest stages of development the zygote divides into 2, 4, 8, 16 and 32 cells successively, and so on, although cell divisions soon get out of step with each other. In this way, a multi-cellular organism is formed and it is a matter of complete indifference whether we speak of the organism's being 'built up from' cells or whether we think of the cell as a sub-division of the organism as a whole: sometimes the one description will be thought apt, and sometimes the other. At all events the nucleus of the cell is that part which contains the DNA and therefore the genetic information — the instructions that specify the nature of the cell's synthetic activities and its other physiological performances. Nuclei usually are roughly central and are surrounded more or less symmetrically by the non-nuclear part of the cell — the 'cytoplasm' — which carries out the cell's executive functions. Thought of in this way, the cellular layout of tissues is obviously a biophysically feasible way of ensuring that the activities of a tissue, and therefore of the body generally, are under the control of nuclei.

The 'Cell Theory' declares that all tissues are cellular in composition or, if like bone they consist very largely of inorganic structural matter, are cellular in origin. The greatest triumph of the Cell Theory was the recognition that contrary to all superficial appearances even the nervous system is cellular in nature. The constituent cell is called a 'neurone', and as might be expected in view of their special function in the propagation of nerve impulses neurones have departed very far from the kind of archetypal globular structure people tend to associate with cells. The most obviously cell-like part of the neurone is called the cell-body or *perikaryon*, and this houses the nucleus. The propagation of nerve impulses is made possible by extremely long filamentous extensions of the cytoplasm which form nerve fibres or 'axons'. The source of information for specifying all the characteristic acts of synthesis carried out by the neurone remains nevertheless within the nucleus of the cell-body, and this raises quite a tricky* problem about how such a control can be exercised at a distance which may, in a large animal, be as much as several feet. What in fact happens is a continuous flow or translation of material from the cell-body down the length of the nerve fibre. When a nerve fibre is cut so that part of it is severed from the cell-body, it simply dies and regeneration takes place from the end still attached to the cell-body. This regeneration is, in effect, a continuation of the process of flow of substance from the cell-body.

Cells differ so very greatly from one another that it is hardly possible to describe anything that might be called a 'typical' cell, either from the point of view of structure or of activity. There is no doubting, though, that the kind of cell that comes to the mind of a cytologist when he tries to envisage a 'typical' cell is a *fibroblast*. Fibroblasts are a race of cells belonging to the connective and skeletal tissues, and they include fibroblasts from the fibrous connective tissues, osteoblasts from bone and chondroblasts from cartilage.

Regardless of all the reservations already expressed or

---

* Tricky because the kinetics of diffusion are such as to rule out completely an ordinary process of diffusion as a means of communication between the cell nucleus and the extreme tip of the nerve fibre.

intimated about the folly of regarding the fibroblast as a typical cell, they nevertheless form a class numerous enough to deserve to have their properties described. When alive and healthy they are almost invisible when inspected through an ordinary light microscope, and it is only when sick that they become refringent and therefore easily visible; but under a microscope specially adapted to the inspection of living cells (the 'phase-contrast' microscope) the nucleus is easily visible as a darker ellipsoidal body and the cell margin, if the cell is attached to a substratum such as glass or mica, will be seen to have formed filamentous or flat extensions known, idiotically enough, as *pseudopodia* (see p.91). Biology abounds in such examples of comically grandiloquent terminology. The extremity of folly is perhaps the use of the term *pseudo-navicella* to describe a reproductive element that occurs transiently in the life cycle of an undistinguished parasite of the earthworm — a term all the more remarkable for the fact that the object to which it applies could not be mistaken for a real small boat even by a zoologist in need of psychiatric attention.

The Victorian zoologists who coined these silly terms must have been trying to prove something, if only to themselves: in an era in which snobbishness was carried to an extreme hardly now intelligible, it could have been an attempt to persuade those who might otherwise look down on them that zoologists were rather deep and knowing fellows, whose science was not to be penetrated by the uninformed laity.

By means of these pseudopodia, combined with active changes of shape of the cell-body as a whole, fibroblasts can glide round their substratum at speeds of the order of milli-metres a day. Other cells, among them most of the white blood corpuscles and especially lymphocytes, can move at a rate of millimetres an hour. When fibroblasts divide, as most cells can (inability to divide, as with neurones, which are agreed to be too highly specialized to be anything but end-cells or red blood corpuscles, which have no nuclei, always requires a special excusatory comment), cellular movement plays an important part in separating the daughter cells. Fibroblasts are not themselves phagocytic, i.e. they cannot envelop and take into themselves small particles such as bacteria, but in common with many other cells they have the power rather mysteriously to imbibe tiny little droplets of whatever medium they may be living in, known as *pinocytosis*. Another structural element that would show up in the 'phase-contrast' microscope is a number of small, agile, spherical or rod-shaped particles known as mitochondria, the seat of respiration within the cell.

The mobility of fibroblasts and their power to divide makes them particularly suitable subjects for *tissue culture*, a process by which tissue fragments are kept alive and, if capable of it, growing, in nutrient media at body temperature in suitable glass vessels or cells. Fibroblasts are ubiquitous and so readily do they adapt themselves to life *in vitro* that they quite often grow in tissue culture even when the person culturing them is under the impression that he is growing something else. The growth of cells in culture has been a great blessing to experimental biology; it has also a certain relevance to the problems of ageing (see below) and has been the making of medically applied virology, because viruses can be propagated in tissue culture and mass-produced to make possible the large-scale manufacture of vaccines. Tissue culture is not a very difficult technique and the nutrient media in which cells grow, being mostly of animal origin, such as blood serum and tissue extracts, are not hard to come by. However, if it is to give reliable and reproducible results it calls for great fastidiousness and attention to detail: bacterial and fungal infections are the enemies of tissue culture and the media in which cells grow suit them admirably, so a strictly aseptic technique is unconditionally necessary. Because of the need to grow viruses for vaccine production and because some kinds of cells, e.g. lymphoblasts, are needed in large quantities for such purposes as the production of antilymphocyte serum, tissue culture has been scaled up from the laboratory equivalent of the cottage industry to an industrial process. When cells are grown in very large numbers they grow as colonies of separate cells in a fluid medium, in contrast to the traditional semi-solid medium used in the early days of tissue culture, which had the advantage of making the cells specially easy to study microscopically. In laboratories, tissue cultures are used nowadays above all else to study problems of cell biology under the specially well-defined conditions made possible by withdrawing a cell from systemic influences, so that its environment can be completely regulated and exactly known; there was, however, a dark period in the history of tissue culture when the phenomenon of cell growth *in vitro* was thought so surprising and its results so beautiful that tissue culture was studied for its own sake, almost as an aesthetic pastime without any truly analytical purpose in mind. Another disincentive to the use of tissue culture for analytical purposes was the fact that in the conventional tissue culture media then in use the medium was held in a semi-gelatinous state by clots of blood fibrin, which did not allow the commingling and close contact between cells which is now known to be necessary for many of the processes tissue culture is used to study.

In the early days of tissue culture it was believed (on the basis of experiments long since agreed to be faulty) that tissue cultures were 'immortal', i.e. that when a growth had reached a certain size it could be divided into two or more parts, each of which would become the starting point of a new culture, which would in turn grow and be sub-divided when necessary into two, so that the cellular lineage was in effect without end. However, the work of Hayflick, confirmed by most of the principal tissue culture laboratories throughout the world, has made it clear that so far from being immortal, tissue cultures of normal cells have a strictly determinate lifespan. They survive for a limited number of cell generations and then die out. According to Hayflick, the length of time they live depends upon the age of the organism from which cells for culture were taken in the first place. Cells that have undergone a malignant transformation, either in the body or as a result of virus infection in the tissue culture itself, are not subject to this rule: they seem to be able to proliferate indefinitely. The experiments upon which the notion of immortality originated were supervised by Dr Alexis Carrell at the Rockefeller Institute in New York, and the culture was supposed to have survived from about 1912 until 1939, but if it did so, it was for reasons which are still not entirely clear. One possibility is that the pabulum on which the cells were fed was a coarse embryonic tissue extract containing an abundance of suspended cells which the acolytes of the cultivation ceremony did not take sufficient pains to remove. An alternative and less creditable possibility is that the cultures *did* die out, and were simply started anew from fresh tissues on the grounds that their death could only have been due to lack of attention, to the use of a toxic medium or to some other such accident.

A number of biologists had the idea that — because of their exuberant growth, lack of differentiation and other superficial resemblances to malignant cells — if tissue cultures were to be reimplanted into the body they would turn out to be cancerous. Upon reimplantation, however, progressive growth did not occur. We now know that the tests were not adequate for their purpose, for in the days before the development of inbred strains of mice the cultures being tested for malignancy would of necessity have been transplanted into mice differing in genetic composition from those from which the cultures were started. Thus the tissue cultures would have aroused and been destroyed by an immunological transplant-rejection process (Chapter 13). When this fault of experimental design was remedied, it soon became clear that quite a number of tissue culture lines *had* undergone a malignant transformation, especially in laboratories where oncogenic (cancer-causing)* viruses like polyoma virus were present. Considered as a whole, these experiments are a lesson in not being too easily discouraged by negative results.

Hayflick's experiments also have an important bearing upon the phenomenon of ageing, which will be discussed later (Chapter 20).

The significance of cell structure. It is now possible to give a fairly confident answer to a problem that very naturally puzzled all early students of cells, particularly those whose thoughts were dominated by the notion of 'protoplasm', the mystic polyphasic colloid referred to on p.9. The outline of the problem was this: a cell performs a number of synthetic and other metabolic activities, and certainly depends for its life upon a process, respiration, which calls

---

* Oncology is the science of tumours and the prefix 'onco' always refers to tumours, e.g. oncovirus, oncogenesis.

for the very nicely integrated co-operation of a series of enzymes. How on earth are all these activities so integrated that the right events take place at the right time and in the right place? In terms of a protoplasmic theory the phenomenon is totally unintelligible: it is no wonder that R.A. Peters put forward the notion of a 'cyto-skeleton'. An agreed answer to this conundrum has been made possible by the use of electron microscopy, particularly by using electron microscopes of intermediate resolving power (see p.20). The answer, then, is that the integrated performance of all the multifarious activities of the cell has a *structural* basis, and so far from being filled with homogenous slimes, whether primordial or of a sophisticated latter-day kind, cells are filled with complicated solid-looking organs such as the nucleus itself and the mitochondria, already referred to above (p.124), which provide the material substratum for the positioning and spacing out of the enzymes which transact the cell's metabolic business. Prominent among these intra-cellular structures, especially in cells engaged in synthesis or *secretion*, is the protoplasmic *reticulum*, a convoluted quasi-tubular internal network closely associated with the tiny round particles, the ribosomes, which are the seats of macro-molecular assembly in the cell. Within the nuclei are smaller objects, *nucleoli* of much the same shape, which are repositories of a form of the nucleic acid which plays an important part in the synthesis of cellular proteins (see Chapter 12). Among other organelles whose molecular structure is quite obviously the basis of function are the chromosomes themselves, of course, being made of DNA combined with a basic protein in a salt-like linkage. In Chapter 3 it is explained that the information-carrying function of the nucleus is made possible by, and is an expression of, the molecular structure of the nucleic acid within the chromosomes. Mitochondria have a characteristically ellipsoidal structure with closely packed transverse internal sub-divisions, *cristae*, that seem to fit them admirably for the ordering of the enzymes that transact cellular respiration. Although most of the constituents of mitochondria are synthesized according to the instructions contained in the nucleus of the cell, mitochondria are nevertheless highly unusual in containing some DNA, which is thus not wholly confined to the nucleus. They are highly mobile cell elements, and are biogenetic, i.e. they form anew only where one has existed before, and they multiply themselves by a process which looks remarkably like simple binary fission. An exciting speculation about the origin of mitochondria, for which there is a good deal of circumstantial evidence, is that they began their evolutionary career as bacteria which, starting as intracellular parasites, have since become obligative symbionts. No one has yet succeeded in cultivating mitochondria outside a cell as one may cultivate most bacteria, but this should not be regarded as fatal to the bacterial theory of their origin, because the enterprise may succeed when the right conditions are hit upon and it should not be forgotten that some bacteria, notably mycobacteria, are also very difficult to grow in cell-free media outside the body.

Most of the organelles so far referred to are easily visible in the cell given the right conditions of microscopy, though it is notorious that chromosomes are not visible in the resting nucleus, i.e. in the nucleus between cell divisions. This has never been a source of anxiety because genetic evidence confirms their continued presence in the nucleus whether the cell is dividing or not. Every now and again, however, some oddly shaped chromosome may protrude from the nuclear membrane and thus make itself visible between divisions, and it is a noteworthy and useful curiosity that the presence of the extra X chromosome (see p.40) often makes the nucleus of female cells recognizable as and when they are prepared for microscopy, and this ability to 'sex' cells can be invaluable in situations in which it is essential to be certain of the provenance of a cell or tissue. Some of the locomotory devices found in protozoa are found also in some of the cells of higher organisms: pseudopodia have already been mentioned (pp.91 and 123) and others are flagella in sperm and cilia in some *epithelia*, e.g. of the trachea (see p.129). Cilia and flagella have essentially the same structure — exquisitely delicate filamentous extensions of the cell membrane — and essentially the same way of working — a self-energized unduloid contractile wave passing down the length from base to tip producing the snake-like movement that may occur in a piece of string or length of leather if we try to 'set up waves in it' by a motion of the hand.* A cilium is so short that it cannot carry more than one wave at a time and, since these waves alternate on the two sides, cilia sometimes give the impression of 'lashing to and fro' — a most inaccurate description of how they work. From a molecular standpoint, the contractile events in cilia, flagella and muscle fibres are closely related. In higher organisms, ciliated cells are very often found in situations in which the function they perform is to waft fluid or sheets of mucus across a surface — e.g. in the trachea and the windpipes generally.

The cells of arthropods are devoid of cilia, for reasons doubtless connected with their manufacture of a tough skeletal material known as chitin — which forms the hard outside, and internal sub-divisions, of insects and crustaceans — even in places where they might have been expected to serve a specially useful purpose, e.g. in the respiratory tubes of insects — *tracheae*.

Epithelia and epithelial tissues. When cells of the same kind and the same orientation are laid out in such a way that they touch each other without the interposition of fibres or cells of other kinds, they are referred to as *epithelia* when they bound convex surfaces, and as *endothelia* when they bound concave surfaces, although sometimes 'epithelium' is used as a general term to cover both kinds. Thus, the outermost layer of the skin and the cornea of the eye is always called epithelium and the layer of cells that forms the inner lining of the cornea — the inner lining, therefore, of the anterior chamber of the eye — is always called an endothelium. There is no great consistency about it, however, because the inner lining of the trachea and of the esophagus is always referred to as an epithelium. These inconsistencies are matters of convention, and no one is confused by them.

Connective tissues. In connective and supportive tissues, the constituent cells — members of the great family of fibroblasts — are separated from each other by the matrices or ground substances they themselves manufacture. The

---

* Cilia and flagella are *not*, however, energized from the base, for the contractile wave does not diminish in amplitude from base to tip.

characteristic skeletal product of the connective tissues that bind and support most of the organs of the body are the extremely stout, unbranched fibres of very high tensile strength called collagen fibres. These are knitted together in a three-dimensional web of extraordinary toughness and resilience in that part of the skin — the 'corium' — from which leather is made, or they may form a two-dimensional lattice work in the tough sheaths that envelop, for example, liver, kidney and guts; alternatively, collagen fibres may be lined up one-dimensionally to form a skeletal structure — tendon — of quite amazing tensile strength. The deposition of collagen is sometimes a response to chronic tension: for example, a nerve trunk that is kept for long periods on the stretch will eventually turn into a passable imitation of tendon, owing to the deposition of collagen fibres along the line of pull. This unfortunate transformation occurs when a divided nerve is repaired by suturing together the severed ends under undue tension.

Skin. Apart from its more obvious mechanical functions, the skin is a major organ of temperature regulation and an important depot of fat storage: in animals such as whales the fatty layer of the skin may be more than a foot thick. Most mammals have a muscular layer in the skin as well, and this makes it possible for them to twitch their coats as dogs do, but in human beings such muscles are confined to the face and neck: they are the muscles of facial expression.

The power of skin to regenerate is as a rule greatly overestimated: its ability to do so in reality is very limited. This applies particularly to the leathery layer of the skin. If this is lost through a burn, an excoriating wound or a destructive infection such as smallpox it does not form anew; on the contrary, the space it once occupied fills up with a spongy-looking and very highly vascular tissue of repair called 'granulation tissue'. In due course this is covered over by the ingrowth of epithelial cells from the outermost layer of the skin — the epidermis — from the margins of the wound, but this too is only a temporary organ of repair, for it does not normally re-form hairs, sweat glands and oil glands. The granulation tissue that lies beneath it forms tough connective tissue fibres, but not in the characteristic three-dimensional pattern of normal skin. Healing, so far as it is possible, takes place by contraction, a forcible drawing together of the edges of the wound through the action of forces not yet properly understood; they are at any rate powerful forces, for they are notorious for causing disfigurement and even disablement through the immobilization of joints and severe interference with muscular action. The healing process is thus a rather discreditable one for which the only excuse is that few selective forces can have favoured its improvement, because in real life animals with such grave wounds would almost certainly have died anyway of bacterial infection.

Given the ineptitude of the natural process of skin healing, it is fortunate that modern surgery can improve so greatly upon it. The surgical remedy is skin *grafting*: a very thin slice of skin is moved from some uninjured part of the body and spread over the area that needs cover. This skin will retain the texture, hair-number and quality of the area from which it was taken, and it will prevent contracture almost completely. The whole procedure may be regarded as a rebuff for people who think that 'nature knows best'. Skin grafting is a most notable improvement on nature.

At first sight, the grafting procedure may seem to create one major wound in order to repair another, but what happens in reality is that the skin of the donor area is sliced off at a mid-level to leave behind enough of the leathery layer of the skin to prevent contracture and to expedite healing. A new surface layer forms by upgrowth of skin epithelium from the bases of the hair shafts.

Hairs are formed by and are encased within little tubular invaginations of the skin surface known as *follicles*. Although there are certain rare exceptions, it can be taken as a general rule that, once lost, hair follicles do not regenerate and that hairs do not increase in number during life. Hairs, like the outermost layer of the skin that surrounds them, owe their colour to the activities of special cells known as *melanocytes* which manufacture and deposit pigment granules. Inasmuch as melanocytes can only arise from previous melanocytes and not by the differentiation anew of some precursor cell, it follows that once a hair follicle has lost its complement of melanocytes for any reason (they are more vulnerable than ordinary skin cells to some forms of injury, including radiation injury) the hair becomes colourless and remains so. Hairs are nourished through a little network of blood capillaries around the hair follicle. The idea that hair growth can be speeded up or changed in quality by medicaments applied to the scalp is based upon an interesting but understandable confusion of thought derived from gardening: flowers and crops are watered and their growth is improved by the application of fertilizers, but the difference is that plants are alive, are held in what is predominantly an inorganic matrix of soil and are normally nourished through their roots, while hairs are dead structures, manufactured and extruded by a living organ, the hair follicle, which is nourished as other organs are by its blood supply. Coloured people have the same number of melanocytes in the same places as whites, so that melanocytes cannot be made the basis of a supposedly anatomical distinction between the two: the difference between them is in the activity of the cells — the melanocytes of coloured people manufacture more pigment. Even so, exposure to sunlight can make dark skin darker still. For this reason coloured people who prize a relatively light skin often avoid exposure to direct sunlight.

All the very different cells in the body are descendants of a single cell — the fertilized egg — which must therefore contain all the genetic information necessary to specify their structure and, moreover, all the information necessary to secure their correct compounding into complex tissues. In French, but no longer in English, this amazingly complex series of transformations is described as an 'evolution'; it is relevant, though, that when Herbert Spencer expounded the theory of evolution he allowed himself to marvel at the unwillingness of his contemporaries to contemplate the idea of an evolutionary origin of species when an equally marvellous process distantly analogous to evolution — the development of the adult from a single cell — was accepted by them as a matter of course.

Chapter Sixteen
Circulatory Systems

Although there are accumulations of fluid in several parts of the body — in, for example, the perivisceral cavity, the cavities of the brain and spinal cord and the anterior chamber of the eye — all such fluids undergo a continuous process of drainage and refreshment: there is no standing water anywhere. The two systems in the body in which fluid is circulated mechanically are the blood system and the lymphatic system.

Everybody knows that arteries are the vessels that conduct blood away from the heart and that immediately by the heart they are very broad, stout pipes whose structure enables them to withstand the strong pressures to which they are continuously subjected. Moving away from the heart the vessels become less stout and their branching angles become wider and wider. Arteries contain circumferentially disposed muscle fibres as well as connective tissue — and so do veins, but there is less muscle in veins. As the arteries get smaller, the pulsatile character of the flow becomes less and less obvious and at the level of capillary vessels the flow is effectively continuous. Capillaries are the final ramifications of blood vessels in the tissues — a closely textured network of minute intercommunicating vessels — the 'capillary bed'. Capillaries (which in spite of their name are very much finer than hairs) come together as venules and these in turn unite into veins which eventually return the blood to the heart. Nobody will be dumbfounded to learn that the heart is the principal circulatory organ of the blood system, but this is by no means the whole story. Because of the very great resistance offered by the capillary bed, taken in conjunction with the moderate viscosity of the blood, the pumping action of the heart is not in general sufficient to circulate the blood to and through the tissues and back to the heart: the job is completed by muscular contraction which tends to flatten out the thin walled veins and force the blood along them. Veins have valves which ensure that the blood so propelled can travel only towards the heart. Thus an important circulatory function is performed by ordinary muscular contraction. So-called 'arterio-venous anastomoses' short-circuit blood from terminal arterioles into venules and thus regulate the volume of blood in the capillary bed; this in turn has an important influence on heat loss. Under circumstances in which the conservation of heat is necessary, arterio-venous anastomoses tend to be open, so diminishing the supply of blood to the capillary bed and reducing loss of heat.

Blushing is an interesting psychosomatic event in which a sudden shut-down of the arterio-venous anastomoses of the face floods the capillaries with the blood that produces the characteristically heightened colour. People who do not believe in psychosomatic events and do not believe that the mind can influence the body by direct nervous pathways should reflect upon blushing, which in people of heightened sensibility can be brought about even by the recollection of an embarrassment of which they have been the subject — as clear an example of the influence of mind over matter as one could wish for. Most parts of the body are not uniquely dependent upon a single arterial supply: a collateral circulation is usually well enough developed to ensure against what would be the otherwise disastrous effect of arterial deprivation. Some arteries, however, are end arteries, i.e. are uniquely responsible for the supply of a certain zone. Here, there is no collateral blood supply and the consequences of arterial deprivation are extremely serious: one such artery is the retinal artery. One rather surprising exception to the general rule that most areas of the body are not uniquely dependent upon a single blood supply is the arterial blood supply of the muscles of the heart, which is provided by the two coronary arteries, between which there are no major interconnections although the whole blood circulatory system depends upon their continued working. Blood supply to the brain is no less important, but here the 'circle of Willis', a ring of inter-communicating arterial vessels in the pathway to the brain, even out the pressure and share the blood supply.

Blood clotting is an interesting and rather puzzling phenomenon. An easy and apparently self-evident explanation of its function is to arrest the loss of blood through injuries to blood vessels, but this can only be half the story. Blood clotting is too slow a process in itself to stop bleeding: it cannot help if blood clots after it has been shed. It is more likely that what stops the flow of blood is the contraction of the muscle fibres in the coats of the small arteries. Blood clotting may then be relied upon to block up the injured vessel, even after the muscular contraction has ceased. The fibrin fibres of a blood clot make an almost ideal scaffolding for regenerative growth.

Blood is a yellowish fluid — often milky in appearance, particularly after a meal — which contains thousands of cells per cubic millimetre in suspension. The fluid is called blood *plasma* and it contains the protein fibrinogen which is responsible for forming the fibrin blood clot. Plasma from which the cells have been removed clots to form a jelly, but when whole blood clots fibrin entrains all the corpuscles within it and when the clot contracts, as it invariably does, it expresses a clear yellow fluid known as *serum* which is no longer capable of clotting; thus, roughly speaking, serum equals plasma minus fibrin. It is serum, not plasma, that is used for most medicinal purposes, e.g. the transfer of anti-bodies when it is necessary (see Chapter 13, Immunology).

In most warm-blooded animals the blood corpuscles are of three kinds: red blood cells (erythrocytes), white blood cells (leucocytes) and 'blood platelets', minute cellular fragments formed from the cytoplasm of other cells, which are incapable of division or of behaving in characteristically cell-like ways. The platelets act as foci in the clotting process and tend to enhance it by their deposition on damaged tissue surfaces. Red blood corpuscles owe their colour to the iron-containing pigment haemoglobin, the reversible oxygenation of which is the basis of oxygen transport throughout the body. Some Antarctic fish have metabolic demands so low that they can be met merely by the oxygen physically dissolved in the blood plasma, but warm-blooded animals are metabolically much more active and none lacks haemoglobin. True white blood corpuscles are of three kinds: polymorphs, monocytes and lymphocytes — all actively as well as passively mobile; the first two are highly phagocytic, and specially well able to engulf bacteria, particularly when the bacteria are covered by antibodies directed against them (see Chapter 13, Immunology). Polymorphs are the cells that form the bulk of pus that accumulates in local infections like boils and for that reason are sometimes known as 'pus cells'. Lymphocytes are the characteristic cells of the lymphatic system and are best dealt with under that heading (see p.137).

Whereas the blood circulatory system goes both to and from tissues, the lymphatic system conducts fluid *from* tissues only, but lymph, the fluid contained within it, also circulates. Lymphatics begin as a closed (i.e. not open-ended) network of interconnecting and blind capillaries within the tissues which unite first into lymph venules and then into larger lymphatic vessels sometimes called 'lymph veins'; the largest lymph veins — the right and left thoracic ducts — eventually open and pour their contents into the venous system. Lymph is a term that has been loosely used by physiologists to describe any fluid that can be expressed from tissue, but strictly speaking it should be used only to refer to the contents of lymphatics. Lymph is derived from and is in equilibrium with blood plasma. It is colourless because it contains no red blood corpuscles, though after a fatty meal lymphatics issuing from the intestinal region look milky, because of the suspension in them of vast numbers of tiny fat droplets. Lymph contains the blood-clotting protein fibrinogen so like plasma it can clot. Lymph derives from blood plasma by a process of filtration, regulated by a balance between blood pressure and the water-retaining power of the plasma conferred upon it by its so-called 'colloid osmotic pressure'. When the blood protein level is gravely depleted, as it is in chronic starvation, the tissues tend to become waterlogged: the 'oedema' that is such a tragic symptom of starvation. The lymphatic system has no active propellant analogous to the heart; instead, like venous blood, it is passively propelled by muscular contraction and other bodily movements. Lymphatic veins have even more valves than blood veins and these ensure that lymph passively propelled along lymphatics by the squeezing of lymph vessels always travels in the right direction, i.e. towards the point of entry into the venous system. *Massage* also propels lymph and every masseur knows empirically that, to reduce swelling and achieve the sense of relief that massage is so well known to bring, the direction of massage must be that which is determined by the valves, i.e. in the general direction of the neck. Obstruction of the lymphatics leads to the kind of tissue swelling known as lymphoedema: the grotesque lymphoedema known as elephantiasis is due to a blockage of the lymphatics by parasites.

On their way to the venous system all lymphatic vessels pass through one or more *lymph nodes*, organs whose structure adapts them particularly well to filter off particulate matter in the lymph and which act as stations in the pathway of the lymphocyte circulation as described below. Lymph nodes are often the site of the first immunological response (Chapter 13) to antigens that enter the body through the tissues.

Lymphocytes are the characteristic cellular element of lymph. They enter the venous system, together with the lymph in which they are suspended, in numbers of the order of 10^{10} (tens of billions) daily. This posed experimental pathologists with a real problem, i.e. not a nature-philosophical problem: where do all the lymphocytes go? It is out of the question that they should be manufactured anew in numbers large enough to make good the supposed wastage, so what on earth happens to them? The answer was discovered by J.L. Gowans: they circulate. Lymphocytes present in the bloodstream enter lymph nodes through special vessels, the post-capillary venules, and from the lymph nodes they pass into the lymphatics that lead away from them into the venous system, and so the cycle is completed. A small minority of lymphocytes may pass through ordinary blood capillaries into ordinary body tissues and find their way into the lymphatic capillaries that start there. In any event the end result is the same: lymphocytes circulate and recirculate, so that the cells present in the blood at any one time are like the chorus of soldiers in a provincial production of the opera *Faust* — they make a brief public appearance and then disappear behind the scenes only to re-enter by the same route. It is not surprising, then, that lymphocytes are long-lived cells — much longer lived than red blood corpuscles. Red blood corpuscles last only a matter of weeks, but lymphocytes live for years, and even ten years is not an unheard-of figure. The long lives of lymphocytes and their capacity to visit every part of the body equip them admirably for their main function, which is to transact immunological responses (Chapter 13). Although lymphocytes look as alike as the English to the Chinese they are in reality sub-divided into two great classes by function and by pathway of development, and not by any obvious visible characteristic. These are B lymphocytes, the derivatives of which manufacture antibodies, and T lymphocytes, which transact cell-mediated immunological responses.*

The *thymus*, a most important lymphoid organ, is a lesson to those naive enough not to believe in teleology — the belief that biological organs' performances evolve and develop in such a way as to fill a certain purpose (although the purpose is not a conscious one and, contrary to what Aristotle might have believed, the purpose cannot be the cause of its evolution or function). The thymus gland, particularly in young mammals, is an astonishingly large, juicy-looking organ which occupies the front end of the thoracic cavity. Many biologists seeing it for the first time must have said to themselves, 'I shouldn't wonder a bit if this organ were found to have some important function,' and those biologists who entertained this audacious thought have been triumphantly vindicated in recent years: the thymus turns out to be a finishing school for a certain class of lymphocytes that originates in the bone marrow — the 'T' lymphocytes that are the agents of cell-mediated immunological responses (Chapter 13, p.100). Congenital absence or experimental removal of the thymus very gravely impairs the maturation

---

* 'T' is short for 'thymus-dependent' and 'B' for the *Bursa of Fabricius,* which in birds has a function analogous to the thymus but exercised on behalf of antibody-forming cells.

of the cell-mediated arm of the immunological response.

In summary, a vertebrate body has not one but *two* circulatory systems, the blood system and the lymphatic system, which intercommunicate and are functionally closely dependent on each other. In the circulation of blood and lymph a critically important part is played by ordinary muscular contraction, which constricts blood and lymph vessels in which the valves ensure that the fluid which passes along them can only go in the one 'right' direction.

Chapter Seventeen

Co-ordination Systems: Hormones and Nerves

The working of the body requires a highly efficient co-ordination of its several activities. In animals an event in one part of the body influences or is influenced by an event elsewhere through the mediation of hormones or of nerves. When teachers introduce their pupils to hormones it is quite common for them to indulge in a little charade of simulated puzzlement about what hormones 'really' are. However, this is not a problem that gives even a moment's anxiety to endocrinologists — the biologists who specialize in the study of hormones and the glands that manufacture them. Hormones are substances such as thyroxine and triiodothyrodine, which are products of the thyroid gland, insulin (produced by that part of the pancreas which does not manufacture digestive enzymes) and cortisone-like substances manufactured by the outer shell of the adrenal gland; all are biological agents that circulate in the blood and exercise either a general or a specific stimulatory or inhibitory effect on biological processes elsewhere in the body. Congenital or accidental deprivation of particular hormones — e.g. of insulin, of adrenal hormones in tubercular disease of the adrenal glands, or of thyroid hormones owing perhaps to a dietary insufficiency of the iodine required to synthesize them — usually gives rise to recognizable diseases or functional imperfections of the body such as diabetes mellitus, excessive loss of salt or cretinism respectively — diseases that can often be remedied by the artificial replacement of the missing hormones.

Hormones are normally liberated directly into the lymphatics or the bloodstream, whereas with most organs the secretory product is conveyed by a duct to the place where it works. A number of our organs have dual functions, producing both a conventional secretion, such as the pancreas with its digestive enzymes, *and* a hormone. Another example is the testis which, in addition to producing spermatozoa, also manufactures male sex hormones.

Deprivation and replacement is the classical experimental method of demonstrating hormonal function in an organ, but obviously this procedure cannot be applied to organs that are essential for life for reasons unconnected with the hormones they manufacture: we can be pretty sure that more hormones exist than have yet been identified.

There are a number of interesting curiosities about the evolution of the hormone-secreting glands; one of them is that such glands are often homologous with organs which performed some quite different function at an earlier stage in evolution and have become functionally redundant. The thyroid gland, for example, is homologous with the apparatus which in the remotest ancestors of vertebrates — animals like sea-squirts — played an important function in trapping minute particles of food from the stream of sea water that flowed through the pharynx or branchial basket. In such a case as this, and very probably in a number of others, it is quite likely that when the principal function of the organ has been lost its hormone-secreting function has simply been retained. The working of the thyroid gland in land vertebrates is very much what we should expect in the homologue of an organ that had special opportunities to accumulate the element iodine, which is essential for the manufacture of its secretory product, and certainly the general regulation of the rate of metabolism in the body (a function of thyroid secretions) is not unexpected in an organ playing an essential part in the throughput of nutritional matter in our remotest ancestors. Amongst the other hormone-secreting glands that are derived from the functionally obsolete branchial basket or 'endopharyngeal apparatus' of organisms such as sea-squirts are the parathyroid gland and the thymus.

A second evolutionary curiosity in the tendency of hormone-secreting glands is that tissues that were at one time scattered and diffuse eventually come together to form a single compact organ. Yet a third is that organs of totally different origins and functions may become structurally involved with each other in the course of evolution — the most famous example being the core and the outer shell of the adrenal gland, the former derived from a modified sympathetic ganglion exercising nerve-like functions (see below) and the outer shell derived from what may be the rudiment of a very primitive kidney-like structure which had to do with the regulation of salt concentration in the body.

Specially important among endocrine glands are those that have other such glands as the targets of their action: thus hormones manufactured by the so-called 'anterior' part of the pituitary gland regulate the working of the thyroid, the gonads and the outer shell of the adrenal gland. Hormones that regulate the secretion of other hormones are known by the suffix 'trophic' so that the thyrotrophic hormone is that which regulates the working of the thyroid gland and gonadotrophic hormones affect the maturation of the tissues that produce the germ cells.

Hormones owe their specificity of action partly to their chemical structure (for ostensibly tiny changes in the molecular structure of steroid hormones have a profound effect upon the way they work) and partly to the particular character of the receptors on the target cells that engage with the hormone.

Endocrinological, i.e. hormone-mediated reactions, have no 'memory' except in the sense in which that word would be used in a neurological or immunological context: for whereas it is a distinctive feature of all immunological reactions that a second exposure to an antigen excites a reaction different from that provoked by the first exposure, a hormone usually exerts the same biological action even after repeated exposure. Upon this principle depends the regulation of dosage of insulin in the treatment of insulin-dependent diabetes.

Hormones and nerves compared. Nerves work quite differently from hormones but there is an extensive area of overlap between hormone-mediated reactions and reactions mediated through nerves. For one thing, some hormone-secreting glands are derived from neural elements — an origin already mentioned in relation to the core of the adrenal gland, but best exemplified by that part of the between-brain known as the hypothalamus, which manufactures not only the hormones once thought to be manufactured by the so-called 'posterior' pituitary but also the trophic hormones ('releasing hormones') that activate the anterior element of the pituitary gland. Then again, some part of neural transmission is itself mediated through the action of hormones. Contact hormones such as acetylcholine act at the very point at which they are manufactured and are then promptly destroyed.

Ductless glands of neural origin exercise very much the effects that might be anticipated from their site of origin; thus the hormone adrenalin or epinephrine of the core (*medulla*) of the adrenal gland exercises just the function that would be expected of an organ derived from a sympathetic ganglion. Again, the endocrine effects of the hypothalamus are just what would be expected of that part of the brain which is the principal centre of so-called autonomic motor functions, i.e. functions that have to do with the regulation of the 'involuntary' muscles that are normally hoped and expected to work on their own without conscious attention.

The specificity of nervous action is, of course, to a very large extent anatomical, i.e. particular episodes of behaviour take place because sense organs and effector organs are interconnected by nerves in an anatomically specific way.

Whereas hormones fulfilling different functions are agents with different chemical structures, nerve impulses are all of a kind, no matter from which sense organs they arise or what muscle they supply. A nerve impulse has a character of a propagated change of state, i.e. during the passage of an impulse no substance actually moves down a nerve fibre. Moreover the energy requirements of nervous conduction are met by the fibre itself: there is no such thing as a dying away down the length of nerve of the effect produced by a stimulation at any one point, e.g. at a sense organ.

Motor co-ordination of the kind that leads to the performance of complex behaviour tends also to be interpreted in terms of the specificity of neurological connections, i.e. of the interconnections made between the nerve cells themselves in the great correlation centres of the brain. So very large is the number of nerve cells in the brain and so very much larger, therefore, is the number of possible interconnections between them that it is difficult to see how such an interpretation could be found wanting in principle. The difficulty is to provide a critical test of it or to imagine by what means it could be disproved if it were indeed faulty. Much offence has, inexplicably, been taken at the notion that the brain works in a manner analogous to a computer — 'inexplicably' because no one takes offence at the idea that the eye may in some respects be likened to an optical instrument, such as a camera with a lens that throws a sharp image upon a photo-sensitive film in some ways analogous to the retina. The eye is not a camera, though: it is more useful to think of a camera as a sort of 'exosomatic' eye which performs certain eye-like functions, just as a wooden leg performs leg-like functions and cutlery performs tooth-like functions. In the same way we can confidently say that a computer performs brain-like functions and is a kind of exosomatic brain, with the implication that an understanding of how computers work may teach us quite a lot about how brains work. The lesson at present falls short, however, of explaining the nature of memory. Although it is widely believed that memory may have a structural basis, no one has yet thought of a plausible theory of structural encoding of neural memory, in the sense in which genetic 'memory' is structurally encoded in DNA.

Chapter Eighteen
Sense Organs

Sense organs are instruments of the kind known to physicists as 'transducers' — they exactly translate or map one wave form or energy form into another. Examples of transducers are a microphone and the pickup of a gramophone, both of which translate mechanical vibrations into minute voltage pulses. Biological sense organs respond to changes in the environment and translate them into the currency of nerve impulses. Each different kind of sense organ is adapted to respond to a different environmental stimulus of a kind specially relevant to the organism's welfare. The *modality* of the sense — whether of sight or sound or touch or pain — is however not a property of the sense organ but is determined centrally (a truth first discerned by the great German physiologist Johannes Müller and embodied in his so-called 'law of specific irritability'). Stimulation of the optic nerve, no matter how it is brought about, causes sensations of light and stimulation of the auditory nerve sensations of sound: a blow on the eye notoriously gives rise to an impression of a flash of light and some diseases of the inner ear have the effect of producing continuous and sometimes incapacitating 'head noises'.

Sense organs respond to the *changes* in the external environment: continuous uniform stimulation usually excites no response — the subject 'adapts'. Some such processes of adaptation are familiar enough in everyday life. If one goes into a room with a uniform smell one soon ceases to be aware of it, just as one ceases to be aware of the ticking of a clock, but there is an important difference between the two forms of unawareness. After adaptation to a smell no effort of mind can enable us to smell it, but with the ticking of a clock an effort of attention will enable us again to hear it. Thus the one process of adaptation is peripheral, and the other central.

Sensory inputs can continue during sleep, except from those few sense organs such as eyes which can be switched off, i.e. closed. The input of auditory stimuli during sleep is made clear by the promptitude with which we can invent a dream scenario to explain away the occurrence of obtrusive but otherwise irrelevant noises like the ringing of bells (including, alas, the alarm clock).

Dreams.   It was an unlucky judgment of Freud's that dreams are 'pathological' phenomena, for the recent work of the Chicago school has shown quite clearly that so far from being a pathological process, dreaming is a regular accompaniment of sleep and indeed a physiological necessity. Periods of dreaming during sleep are marked by the occurrence of 'rapid eye movements' (R.E.M.). The existence of R.E.M. episodes during sleep has made it possible to experiment on volunteers by depriving them of dreams. It turns out that people deprived of dreams become irritable and even to a mild degree psychotic. The physiological purpose of dreams — apart from the innocent purpose described above (an adaptation to prevent our being woken up by irrelevant sensory signals) — is unknown.

Paired sense organs.   It will not come as a blinding revelation to most readers to learn that distance receptors tend to be paired.

Whatever dreams may be 'for', the purpose of the pairing of sense organs is obvious enough. The pairing of eyes is the property that makes stereoscopic vision possible and the pairing of ears makes it possible to localize the source of sound, for the sounds reaching the two ears will differ in amplitude and in phase. The single median unpaired eye of vertebrates — the pineal eye — has long since become obsolete and like most other organs which have become redundant in the course of evolution it has evolved into something else. It makes, nevertheless, a transient appearance during development.

It is not only the modality of the senses that is centrally

determined, but also the nature and quality of what is sensed. Specific visual receptors may be responsible for perceiving an upright straight line, a horizontal straight line or a straight line that slopes one way or the other. These capabilities are 'programmed' in a way that gives modern sensory physiology a curiously Kantian colour. It was Kant's brilliant conception that both space and time were 'forms of intuition', i.e. were imposed upon a real world by the character of the intuitive process. In the light of modern sensory physiology Kant's ideas no longer sound as extravagant as they once did.

It has been complained of Kant that he was unaware of a sixth sense, *proprioception*, which might have caused him to modify his views on spatial intuition if he had been aware of it. This is unlikely, because that part of the philosophy of mind which deals with sensation is so very largely dominated by visual sensation that even the recognition of proprioception or pain as modalities of sensation would probably not have influenced Kant very greatly. Proprioception is that form of sensory self-awareness that gives us knowledge of our own posture — knowledge, for example, that the knees are bent even if we do not see them.

It is very doubtful if any of the evidence amassed by modern sensory physiology would have been thought relevant by Bishop George Berkeley to weighing up the merits of 'realism' and philosophic idealism. He would probably have pointed out that all evidence relevant to making a choice between them was already to hand and that inasmuch as all our knowledge of the working of the senses reached us through sensory pathways no accumulation of such knowledge, however voluminous, could draw aside the veil between the world of ideas and the world of reality. As with many other problems that are in principle incapable of solution, this one has been dealt with by being disregarded altogether, except as an exercise in dialectics or as a subject for disputation very suitable for students of philosophy.

Chapter Nineteen

Animal Behaviour and Human Behaviour

The study of animal behaviour went through a bad period during what might be called the exultant era of mechanistic thinking in biology — an era still remembered in the obsolescent terminology of 'tropisms' and 'taxes'. During this period, the organism was treated as if it were a sort of puppet or marionette, worked by strings which represented so many sensory inputs and by internal cross-connections and pulleys which defined completely the possibilities of the organism's behaviour.

The barrenness of this method of approach, and its failure to provide any satisfactory understanding of how animals behave, led to its being supplanted — under the influence of Konrad Lorenz and Niko Tinbergen — by the science now everywhere called *ethology*.

The older students of animal behaviour were in thrall to the notion of Baconian experimentation. They felt that it was at all costs necessary to poke or prick animals to see what they did or to confront them with new situations to see how they reacted. Ethology, the study of natural behaviour, was in effect dismissed by them as an outdoor pastime.

Ethology, of which Julian Huxley and the Howards were also pioneers, is a skilful occupation requiring great patience, concentration and a more than average degree of imaginative insight. An ethological exercise begins with the observation of natural behaviour and completes its first stage when performances which to an untrained observer might seem disconnected or pointless are seen to fall into a functionally significant pattern.

The greatest single weakness of the mechanistic approach

to behavioural analysis was its failure to give due weight to 'instinctive' behaviour. Ethology has remedied this shortcoming by revealing in one context after another the pervasiveness of the instinctive element in behaviour. Although there is a broad distinction between programmatic and learned behaviour, it is no longer thought possible to draw a sharp distinction between the two. The most detailed and careful ethological studies, such as those of W.H. Thorpe on birdsong — in which the idea of a chromosomal tape-recording is specially appealing — show that even where a programmatic element of behaviour is unmistakeably large, environmental clues and cues are necessary for its complete realization. Even so, the programmatic element in behaviour is large enough and distinct enough to unseat the fundamental axiom of empiricism that nothing enters the mind except by way of the senses: *Nihil in intellectu quod non prius in sensu.*

Thanks to the popular writings of Konrad Lorenz, some of the technical concepts of ethology are beginning to pass into vernacular usage, e.g. the notion of 'imprinting', the process by which a newly hatched duckling or baby seal accepts as a mother figure the first object it sees.

If the fault of mechanistic behavioural analysis was its failure to give due weight to the programmatic element in behaviour, the principal fault of modern ethology is that it does not lend itself sufficiently easily to analysis in physiological terms.

From the layman's point of view the most interesting aspect of animal behaviour lies in the light it may possibly throw on the behaviour of human beings.

There seems at first no *a priori* reason why the behaviour of animals should throw any light on the matter: a distinctive and most important characteristic of human beings behaviourally is their possession and exercise of a moral sense. Although many attempts have been made to 'explain' the evolution of this moral sense, no one has commanded universal respect. All of them turn upon the selective value of behaviour that is altruistic and co-operative. Even, however, if a satisfactory theory were to be propounded, its existence would not in any way diminish the importance of the moral sense nor the importance of the distinction between animals that do and do not possess it.

Unfortunately, some attempts to 'interpret' the behaviour of human beings — particularly the more disagreeable aspects of it — in terms of the behaviour of animals has been marred by an incomplete understanding of the animal model from which the lesson is to be drawn. Thus, before interpreting human warfare and internecine aggression as a manifestation of territoriality, as it has been described in birds and other animals, it should be recognized that much of the aggression and protective behaviour of animals in defending their territories is bluff.

Even if certain parallels drawn between the behaviour of human beings and of animals were shown to be unsound, human behaviour has evolved and therefore it would be extremely surprising if *nothing* could be learnt about it from studying the behaviour of animals. For example, the entire complex scenario of sexual behaviour in man — mating behaviour, childbirth and the care and rearing of the young — cannot have sprung into being fully fashioned during the evolution of mankind. It takes no great stretch of the imagination to see that maternally protective behaviour and the suckling of the young in apes and in man are homologous processes: we can learn quite a lot about maternal care in apes by studying it in human beings, and this also holds good in reverse. The work of the Hindes, for example, on the effect of maternal deprivation in rhesus monkey babies is clearly applicable to human beings and it is very much to be hoped that Tinbergen's ethological interpretation of autism in human beings will prove correct and will lead to usable therapeutic procedures. What human psychology has to learn from ethology is a method and a way of going about things — indeed, the more eclectic modern psychiatrists are already beginning to speak about ethology as if they had invented it.

We can expect hints and clues and guidance as to psychological method from the study of the behaviour of animals, but in the future as in the past it is likely that almost all we learn about the behaviour of human beings will be learnt by studying human beings themselves.

The existence of behavioural homologies makes it easy to understand that such human activities as play, showing off and sexual rivalry are not innovations of mankind but have deep evolutionary roots.   The concluding paragraph of Darwin's *The Descent of Man* sums the matter up admirably (though to make sense of it in this context we must substitute 'behavioural repertoire' for 'bodily frame').   The passage runs thus:

> But we are not here concerned with hopes or fears, only with the truth as far as our reason permits us to discover it; and I have given the evidence to the best of my ability.   We must, however, acknowledge, as it seems to me, that man with all his noble qualities, with sympathy which feels for the most debased, with benevolence which extends not only to other men but to the humblest living creature, with his godlike intellect which has penetrated into the movements and constitution of the solar system — with all these exalted powers — Man still bears in his bodily frame the indelible stamp of his lowly origin.

Chapter Twenty

Senescence

Among the most prominent — indeed, loved and revered — characters in sentimental animal literature are those wise and kindly old grandfather and grandmother figures who usually wear pince nez if male and bonnets if female, and get around with the help of a walking stick.

These wise old dotard animals are purely imaginary of course: the work of field naturalists — above all that of David Lack on wild birds — has shown that wild animals simply do not live long enough to give evidence of senile decay. Nevertheless, an experiment may be done which shows that virtually all higher animals can display that deterioration of the faculties with increasing age that is referred to as 'senescence'. The animal is removed from its natural surroundings, secured from its natural predators and provided with food and shelter. Under these highly unnatural conditions it will eventually 'age' purely and simply because it is allowed to live long enough to be able to do so.

In precisely the limited sense of this description, it is true to describe senescence as an 'artefact of domestication'.

This proposition has been taken by people who should know better to imply that domestication is the cause of ageing. It is of course no more the cause of ageing than the amenities of civilized life are the cause of cancer, but it just so happens that in the industrially advanced western countries people live long enough to contract those forms of cancer which rather characteristically afflict the middle aged and the elderly. In countries where the mean expectation of life at birth is only thirty or forty, the incidence of these forms of cancer is naturally very much lower.

The cruel truth about the force of mortality among wild animals, and particularly wild birds, should be studied with special attention by those wont to indulge in the idiotic cry that a caged bird should be 'given its freedom'. The notion of freeing a budgerigar to pay morning calls in the territories of male birds in its neighbourhood cannot but make the blood run cold: the mean expectation of life of a budgerigar freed in this way could not be much more than twenty minutes.

Senescence is a puzzling phenomenon and except in the light of the evolutionary theory discussed below its teleology is most obscure.

Ageing as it occurs in an individual is accompanied by slow, orderly deterioration of all the faculties. Very many hours of profitless labour have been spent in describing and annotating in ever greater detail the changes that accompany ageing in each tissue and organ system of the body.

*Theories of ageing.** The process of ageing is so orderly and predictable that one cannot resist the feeling that it is as deeply programmatic in character as development itself. Although some degenerative changes must depend on others — it has been said that a carnivore is as old as its teeth — the general impression is that senescent changes occur independently in all the organ systems. It is, however, not yet possible to say so with complete confidence, because enough use has not been made of the proposal that *heterochronic* transplantation should be used extensively in the analysis of ageing. Heterochronic transplantation is the transplantation of tissues between animals of different ages — for example, an old piece of skin or an old kidney into a young body or *vice versa*. Such transplantations must of course be done with animals of the same genetic make-up — 'syngeneic' animals.

It was at one time thought that ageing was an 'organismic' property — something that might be expected of a whole organism but not of its constituent parts — but this view is now generally discredited. It was founded upon the belief (already discussed on pp.124–5) that cell lineages such as

* Theories of the ageing process have been discussed by P.B. Medawar in *The Uniqueness of the Individual* (Methuen, London, 1957) and by A. Comfort in *The Biology of Senescence* (Routledge and Kegan Paul, London, 1956).

tissue cultures would grow indefinitely outside the body, but the work of Hayflick, since confirmed by many others, has shown that this is not the case. Tissue cultures have a determinate lifespan which is shorter the older the animal from which the cells to be cultivated have been derived. There is no convincing evidence, however, that the process which determines the life of a tissue culture is closely enough related to the ageing of whole organisms to be informative.

The extreme orderliness, predictability and apparently 'programmatic' character of senescence has been a challenge to biological theorists for very many years. The most famous pathological theory of ageing — the theory that treats ageing as a disease-like process — was that of the Russian zoologist Elie Metchnikoff, a man famous also for the discovery of phagocytosis and for being the first to attempt to devise an antilymphocyte serum. Ageing, he believed, is due to cumulative self-poisoning — autointoxication — by the toxins of bacteria normally resident in the gut. He thought that these processes might be diminished or annulled by replacing the gut flora with lactic acid bacilli. It is to the advocacy of Metchnikoff that we owe the widespread popularity of yoghurt today as an article of diet; and from him also arose the belief that Bulgarian peasants owe their longevity to the regular consumption of sour milk curds.*

Although Metchnikoff's special theory of the ageing process has not found favour, he deserves great credit for having been the first to treat ageing as an epiphenomenon of life — something superimposed upon the normal processes of living — rather than as a phenomenon somehow entailed by the life processes themselves. This is now widely thought to have been a sign of good judgment on his part.

Another theory, Leslie Orgel's, regards ageing as the consequence of accumulated errors of information-processing in the body. Among dividing cells such errors may of course arise in the nucleic acid information source itself, but in addition mistakes are bound to occur in the transcription

---

* Cynics have observed that the consumption of many modern commercial yoghurts with their preservatives and sometimes ill-defined additives would just as surely have brought the lives of Bulgarian peasants to an untimely end.

and translation of genetic information into bodily constituents, and if these happen to be enzymes then the products whose manufacture they make possible will be awry as well. The attempts that have been made by Holliday and others to test this theory — e.g. by feeding organisms on kinky amino acids etc. — have on the whole tended to support it.

Cures of ageing. From a therapeutic point of view, Orgel's theory is a bit depressing because if ageing is due to the process he envisages then nothing very much can be done about it. In general, however, there is no reason in principle why some of the more disagreeable changes that accompany ageing should not be diminished or annulled — particularly those which are clearly secondary in nature, i.e. consequent upon a failure or a deterioration of some other organ system. These are physical changes, no different in principle from other disease processes and no less amenable to investigation and treatment.

The idea of prolonging life has nevertheless aroused a great deal of disquiet among those who are specially conscious of the consequences of overpopulation and the danger of interfering with the natural order of things. It should be borne in mind, however, that almost all advances in medicine may prolong life and it is not at all easy to draw a sharp dividing line between research which is demographically acceptable and that which represents a gratuitous interference with the natural order. As with all life-saving and life-prolonging measures, the real mischief arises from their being adopted piecemeal and haphazardly instead of being part of a well thought-out political programme. By far the most important single factor that has contributed to the growth of populations has been the reduction of mortality in infancy and childhood; there would be no objection at all to the measures that have made this possible if only they had been accompanied by an equal concern over the possible mischief of a disproportionately high birth rate.

The mechanism of a possible evolution of senescence can only be understood in the light of the principles that underlie the *measurement* of ageing.

It is clear that ageing could be measured on an individual and personal basis, by choosing some suitable biological performance and noting how it fell away with advancing years, but the shortcomings of such an approach are so obvious as hardly to need discussion. A measure that has universally been agreed upon also has its shortcomings: it is a measure of *vulnerability* and its rationale is that, inasmuch as no one can die merely from getting a good chronological score, the essential change that takes place in the course of life — to which every source of deterioration will contribute — is an increase in vulnerability, i.e. in the likelihood of dying.

The principle of the measurement of vulnerability is extremely simple: if all organisms die from accidental causes only, e.g. from infection, predation, starvation or accident, and if there were no senescence, i.e. no deterioration with increasing years, then the organism will be no more likely to die when chronologically older than when chronologically younger. Translated into human terms this would mean that a human being would be no more likely to die in his sixth decade than in his fourth or his third or his second.

Imagine, then, a group of a thousand organisms marked or somehow identified at birth and followed through life until all of them have died (*ex hypothesi* from accidental causes). If there is no process of ageing then a constant fraction of this population will die during each year — say five per cent. If the rate of mortality is indeed five per cent per annum the numbers left in the population at the end of each year will be as follows: (1000), 950, 903, 858, 815, 774, 736, 699, 664 . . . etc., approaching zero asymptotically, i.e. getting nearer and nearer but never quite getting there in theory — although it does in real life.

If this series of numbers is plotted against age, it will form a curve of a kind very familiar in physics and chemistry — the so-called 'die-away exponential' curve. As with the curve of unrestricted exponential growth (p.66), the logarithm of the numbers remaining in the population will form a straight line when plotted against time, because it is characteristic of logarithmic graphs that equal distances represent equal multiples or sub-multiples instead of equal increments or decrements, as in the ordinary arithmetic plotting.

This at once suggests a method by which we can ascertain

the degree to which senescence plays any part in the mortality of this population: if the curve of mortality is more-than-logarithmic, it signifies that there is an ageing process, i.e. that vulnerability is increasing with advancing years — the property for which a measure was sought.

Nevertheless, this measure is full of difficulties that should not be underestimated. It takes no account of certain good reasons why an animal's vulnerability should diminish with increasing years. For one thing, animals become behaviourally wiser — wiser in all the senses embodied in 'once bitten, twice shy'. They also become immunologically wiser, for an animal which has been exposed to an infection once and has survived it is thereupon better equipped to cope with the same infection if exposed to it a second time. Then again, some changes that are unmistakeably senescent in character — e.g. the menopause in women — are not accompanied by any increase in vulnerability. On the contrary, some causes of death — such as those associated with childbirth — have now been outlived.

Another objection to the use of vulnerability as a measure of senescence is that it is essentially statistical in character and cannot be applied to a single individual. Nevertheless, if a supposed 'cure' for ageing or anyhow for senile decay is to be tried out on, for example, a population of mice, the only measure of its efficacy that scientists will accept is the degree to which it changes the life table of the mice from its ordinary senescent form into something more closely approaching the exponential die-away form of a population in which senescent changes do not occur.

The orderliness, predictability and ostensibly programmatic character of senescence has tempted many biologists to believe that senescence has evolved in higher organisms; but what could be the advantage of such a process? Weismann thought of ageing as an agency of population control: it eliminated the unfit from the population and so made room and gave extra food to the young and vigorous. It is, however, difficult to see how this would work in terms of an evolutionary process, for if the chronologically older animals were indeed decrepit and generally speaking in the way, the supposedly young and vigorous should have no difficulty in elbowing them aside from the trough and would not in reality be put to a great disadvantage. Weismann was a shrewd man, however, and he may intuitively have suspected that populations do better when propagated through relatively young germ cells — particularly oocytes — than through older germ cells. This is certainly a possibility, but no one has yet demonstrated that any mammal does better for being propagated from generation to generation through young oocytes (in effect young mothers) than through old oocytes — something that would make an interesting subject of research. A modern Galton might, for example, compare the fertility and general prowess of families propagated through older parents with that of families propagated through younger parents. For reasons explained on p.73, in relation to the segregation of oocytes at birth, it is the age of the female patient that is likely to carry the more weight.

Medawar and Williams independently propounded the idea that an evolution of ageing has come about not so much because of any positive advantage in the process of senescence as because of the absence of any disadvantage. Their argument turns upon the fact that, even in the absence of a process of senescence, the numerical contribution that chronologically older animals can make to the population of the future becomes progressively smaller, so that the force of natural selection is progressively attenuated. This is not because the chronologically older members of the population need be supposed to be less fertile, but simply because the forces of mortality acting upon them are so fierce that as time goes on fewer and fewer are left. This will be true even of a theoretically immortal population exposed to a real mortality, for example of ten per cent per annum — an unnaturally low figure. The life table of such a population will follow that of the exponential die-away curve already described on p.156. One consequence of this is that selection in favour of or against genes that come to expression somewhat late in life will be very much reduced, and indeed one method of eliminating a 'bad' gene from a population is to postpone its time of action until the latter period of life. Conversely, it can be shown that if there is genetic variation in the chronological age at which a gene comes to outward expression, then the time of expression of a gene conferring selective advantage will be progressively brought forward in life, while that of a deleterious gene will be progressively postponed; a genetic disaster that befell an organism only at an age which in fact it never reached could not affect its wellbeing in any way. Once begun, any senescence, i.e. deterioration with increasing chronological age, would be a self-reinforcing process. The process of senescence being already established in human beings, the force of selection against any genetic disabilities that occur late in life is reduced virtually to zero.

This 'genetic dustbin' theory of senescence almost certainly represents part of the truth, but equally certainly is not the whole truth.

Whatever the truth may turn out to be, it is now certain that senescence is part of the natural order of things. It is not one that should be interfered with without the utmost care and circumspection. In any case the present incentives to do research into ageing are not very great. The time span of the research is too long for those anxious to make a reputation in a hurry. Many people are rightly deterred by the thought of the possible unwisdom of what they might be doing if they were to be successful. Those anxious about the possible malefactions of research on ageing should take comfort from the fact that the great public and private agencies for research are not competing with each other in their endeavours to support research on ageing, which is in any case a pretty costly business.

Chapter Twenty-One

The Biological Time Bomb Exploded

Shortly after the end of the Second World War, nuclear physicists incurred a fair measure of public odium because of the widespread belief that they were busy devising ever more expeditious ways of exterminating the human race. It was at the same time widely believed (especially by biologists) that biologists were splendid fellows intent upon improving the lot of mankind and busy with all manner of good works.

Unhappily, the frown of public censure has now turned upon biologists themselves: there is scary talk of keeping people alive *ad lib* in the deep freeze, of test-tube babies, of human beings put together from transplants and of a not necessarily benignant manipulation of human heredity.

Although, as will be shown below, many of the grounds for feeling fearful are illusory, the fears themselves are not. *The Biological Time Bomb* was the title given by a well-known science writer, Mr Gordon Rattray Taylor, to a compendium of horror stories of much the kind mentioned above.

Most people will be reassured to learn that the — as we think — idiotic enterprise of preserving human life in the deep freeze is very far beyond our technological capabilities. The same applies to rearing the human fetus outside the body. Birds' eggs develop outside the body anyway, but even today no one has yet reared a bird in an artificial medium from conception until a stage equivalent to hatching. How much less likely, therefore, that such an enterprise should succeed with mammals, whose developmental arrangements are so very much more complicated than a chicken's. However, the fertilization of the human egg *in vitro* and its reimplantation into a prepared uterus is not inherently a very difficult feat, nor is it — in our opinion — an inherently objectionable one: the process may be thought of both as a substitute for, and in many ways a great improvement on, ordinary adoption. It is an improvement because even if the mother has not conceived her child she will have carried it and given birth to it, and thus be physically and perhaps emotionally better prepared for motherhood than someone who adopts a baby 'ready made'.

'Cloning' is another procedure whose possibilities have been widely discussed: cloning is in effect the repeated twinning of a human fertilized egg under test-tube conditions so as to produce an indefinite number of replicas of the original fertilized egg's genetic type. If it is to survive, each such egg must be reimplanted into the uterus of someone prepared to receive it. It is theoretically possible and — on the testimony of frogs and newts — practically feasible to choose the genetic make-up of the egg that is to be repeatedly twinned in this way. Its possibility in principle depends upon the fact that the nuclei of cells in the body that are still capable of dividing — as skin cells are, and lymphocytes — have the same genetic make-up, and could thus, in theory, be substituted for the original nucleus of a fertilized egg which would itself be extruded or inactivated by a micro laser beam. However, even if these procedures were workable in practice and agreed to by all the parties whose consent would be necessary, the problem would still remain — who is to be judged worthy of indefinite replication? The more one thinks of the enterprise the less likely does it seem that it will ever come to pass.

At present, genetic engineering in higher organisms does not run to making directional genetic changes, i.e. to making changes with a specific intended effect.

Some of the biologists involved, nervously aware that many of their proposals are a gross affront to all sense of the fitness of things, have tried shamefacedly to excuse themselves by referring to the importance of genetic engineering in repairing defects of the human genome, such as those responsible for inborn errors of metabolism. Good intentions, however, would not make these procedures any easier.

Although the timorous may be reassured by learning that many of these biological threats are based upon tall stories, they have reason still to be fearful of the Baconian ambition to 'effect all things possible', especially in conjunction with the clear lesson of modern science-based technology that everything which is physically possible will be done if there is a sufficient will to do it. Almost everybody understands that the descent upon the surface of the moon would not have been possible but for political decisions appropriating vast sums of money to the enterprise and attracting an army of experts into its execution.

The carrying through of any one of the bioengineering enterprises described above depends equally upon political decisions, using the word 'political' to mean decisions that would depend upon consensus or concerted judgment of administrative or legislative bodies. Enormous sums of money and tens of thousands of man hours would be necessary to put any one of them into effect, and probably the most difficult part would be to find a grant-giving body gullible or wealthy enough to fund them. Moreover, when they read such scary stories in the newspapers the timorous should remember the deep-seated desire of many academics 'pour épater le bourgeois' or in general to shock the middle classes out of their complacent slumbers; moreover, hunger for notoriety is very often a stronger motive behind many of these stories than a hunger for truth.

Laymen do not realize the width of the gap between conception and execution in science, because they have been misled by a particular kind of fiction in which a young medical scientist, with a dedicated expression on his face, tiptoes purposefully out of the ward to devise, or perhaps even to discover, some new serum or nostrum he has just thought of. It is the width of this gap that makes possible the interposition of wiser counsels and restraining hands between far-fetched ideas and the attempt to put them into effect.

Chapter Twenty-Two

Reducibility and Emergence

The subject of this chapter is a philosophically and biologically live issue. Everybody who studies biology soon becomes aware of the enthusiastic assent or the anxious disquiet aroused by the declaration that biology can be 'interpreted in terms of physics and chemistry'. It is the sense and the implications of statements of this kind that we propose to examine in this chapter. The clearest pronouncement on this principle of reducibility was made by John Stuart Mill in *System of Logic* (London, 1843):

> The laws of the phenomena of society are and can be nothing but the laws of the actions and passions of human beings gathered together in the social state... Human beings in society have no properties but those which are derived from, and may be resolved into, the laws of the nature of individual men.

Applied to physics and biology, Mill's statement would run, 'The laws of the phenomenon of biology are and can be nothing but the laws of the behaviour of atoms and molecules when interacting together in such a way as to form living organisms'.

Both these statements are equally objectionable in their respective ways, and there is a question-begging element in both of them: they might be taken to import that if we knew all about the physiology and behaviour of individual men, we should of necessity know all about society, or alternatively if we knew all about physics and chemistry we should know all about living organisms. Both statements can easily be faulted. There is nothing in the biology syllabus about two-party government or the U.S. Constitution; likewise there is nothing in the syllabus of physics and chemistry, which is already grievously overcrowded, about heredity, infection, sexuality and fear. This is not to be wondered at. Societies are very special kinds of assemblages of human beings and organisms are very special assemblages of molecules. The farthest we can go is to say that sociology is a very special and restricted sub-department of biology. In explaining the nature of these restrictions and confinements on the interactions of organisms we are in effect defining the science of sociology. Likewise, if a physicist or chemist had to think of a name for the science that deals with the very special and highly restrictive forms of interaction between atoms and molecules that can compound them into a living organism, the name he would very likely hit upon is 'biology'.

There is no diminishment here, but simply a recognition that societies are indeed made of individual organisms and individual organisms are composed of constituents which in their turn are made up from atoms and molecules.

If we write down the hierarchy of empirical sciences in the order:

ecology/sociology
organismic biology
chemistry
physics

we can see that each science in the hierarchic table is in some sense a special case of the one below it and that any statement which is 'true' and 'makes sense' in any one of these sciences is also true in and makes sense in any science above it. Everything that is true in physics of the behaviour of atoms is true also in chemistry, and everything that is true in physics and chemistry is also true in biology and sociology; likewise, no matter how specialized the interactions of human beings in forming a society, 'laws' of biology still obtain: human beings still need to eat and breathe and sleep and procreate. To say that a law is true in an upper level science is not necessarily to say that it is interesting or important. A generalization that is as true of sociology as it is of physics is that many of the elements exist in a variety of isotopic forms. 'What of it?' the sociologist or politician may ask. One can only answer that this is the kind of truth that *can* become important in certain contexts — e.g. the fissile properties of certain uranium isotopes can change the history of the world. Each science contains not only the informational content of the sciences below it in the hierarchy but contains specialized notions of its own which do not appear at all at lower levels. Some of these have already been indicated: there is simply no sense in saying that politico-sociological concepts like electoral reform and the foreign exchange deficit can be 'interpreted in terms of' biology, and it is hardly less than idiotic to say they can be interpreted in terms of physics and chemistry, though if the axiom of reducibility were true it would follow that they were so, for the relationship 'interpretable in terms of' is of the kind logicians describe as 'transitive'.* As we go up the hierarchy outlined on p.164, we find that the information content and empirical richness of the sciences progressively increase — as is to be expected — partly because each science contains the theorems of the sciences below it and partly because the restrictions progressively imposed on possible interactions between constituent parts bring it about that each higher-level subject contains ideas and conceptions peculiar to itself. These are the 'emergent' properties. Some analytically-minded people think it treason to admit the existence in any science or at any level of discourse of any property that is not explicable in terms of a science of a deeper analytical level. We believe these fears are quite groundless and that the notion of emergence can be accepted as a straightforward recognition of the formal properties of a hierarchy such as that which is defined by the conventional sciences.** The great appeal of reductive analysis and the

---

* If A is interpretable in terms of B and B in terms of C, then A is interpretable in terms of C.

** There is a fairly close parallel between the hierarchy of empirical sciences and the hierarchy of the geometries according to Felix Klein's conception: see P.B. Medawar, 'A Geometric Model of Reducibility and Emergence' in *Studies in the Philosophy of Biology*, eds F.J. Ayala and T. Dobzhansky (Macmillan, London, 1974).

reason for its almost universal adoption by scientists is that all scientists agree that the actual is most easily understood as a special case of the possible and feel that their understanding of an organized entity A is best secured by finding out how its constituent parts are so ordered with respect to each other that they form A and not B or C. This is also the form of understanding of the world which makes it easiest to see how, if need be, the world may be changed.

For these reasons reductive analysis is the most successful explanatory technique that has ever been used in science. Inasmuch as its virtues and its demerits (which have not been concealed) turn upon purely logical considerations there is no real ground for the fearfulness its use has aroused in the minds of so many laymen. Holists, for example, fear that by analysing the 'organism as a whole' we do it some secret mischief, as a result of which it ceases to be a whole and becomes a mere assembly of its constituent parts. Many earnest but simple people believe that by analysing the song of a bird we somehow make it less joyous and tuneful; again, it was an important element of the literary propaganda of the Romantic Revival that Newton's prism was part of a vast international scientific conspiracy to deprive the rainbow of its wonder and beauty. Keats's denunciation of Newton for destroying all the beauty of the rainbow by reducing it to its prismatic colours had Wordsworth's approval and is recounted in Benjamin Robert Haydon's Autobiography (ed. T. Taylor, vol. 1, 2nd edition, p.385).*

* To which my attention was called by Dr John R. Philip, F.R.S.

Chapter Twenty-Three

God and the Geneticists

Not so very many years ago people talked about 'God and the physicists' — and the names of James Jeans and A.S. Eddington come at once to mind — but today the geneticists have elbowed their way to the footlights and a great change has come about in relations between science and religion: the physicists were in the main very well disposed towards God, but the geneticists are not.

It is upon the notion of *randomness* that geneticists have based their case against a benevolent or malevolent deity and against there being any overall purpose or design in nature.

Randomness enters into the genetic process at two levels: firstly in the entirely random process of mutation, which as already explained plays an important part in providing a candidature for evolutionary change, and secondly in the random allocation of chromosomes to germ cells and the random pairing of germ cells to form a fertilized egg. Indeed, the simple segregation ratios that represent the numbers of offspring of each genetic type expected according to Mendelian rules are quite widely used to illustrate the practical applications of probability theory. It is like a vast lottery in which booby prizes are more obtrusive than rewards. If two parents are both carriers of that deleterious recessive gene which, when inherited from both parents, causes phenylketonuria, then we can pretty confidently say that on the average one quarter of their children will be victims of phenylketonuria, an 'inborn error of metabolism' that may lead to serious mental retardation.

It is not quite good enough to dismiss this unhappy conjunction of deleterious genes as 'bad luck' in the sense in which such a description might be thought to apply to a young parent's being killed by a falling roof tile.

The difference is that a parent's being killed by a falling roof tile presents only the casual and in general terms unforseeable intersection of two otherwise unconnected trains of events, whereas the randomness that enters into genetic mutation and segregation is an integral and essential part of the genetic process, something provided for by the genetic mechanisms themselves. The randomness and the calamities that may go with it are 'laid on' rather than casual.

A disputant intent upon defending the Argument from Design might at this stage declare that the entire genetic system showed clear evidence of design in the way in which Nature provides for and takes advantage of the random element in the genetic process. This is a poor argument and one that would apply almost without qualification to the proprietor of a casino.

Others prefer to see evidence of design in the evolutionary process itself and especially in the characteristic which, wise after the event, is called progressiveness. Such indeed is the theme of the enthusiastic but largely incoherent rhapsody of Teilhard de Chardin *The Phenomenon of Man*.

It cannot be denied that there is an overall progressive tendency in the course of evolution — i.e. that living organisms seem to find ever more complicated and elaborate solutions of the problem of remaining alive and combating a hostile environment. On the other hand, it must not be forgotten that the appearance of progressiveness has in it some of the elements of an optical illusion. We are judging retrospectively when we say that evolution has been progressive in character and in doing so we tend to remember the success stories and forget the failures.

In spite of these reservations, I think it would be unwise to dismiss altogether the progressive tendency of evolution as a phenomenon that calls for no interpretation — it would be unwise, indeed, to assume that our prevailing ideas about evolution were completely satisfactory and left nothing to be explained.

Nevertheless, evolutionary progression and the coming into being of ever 'higher' evolutionary products until it all ends up in what Teilhard called 'a paroxysm of harmonized complexity' will simply not do as a general statement of the direction of the flow of events in nature or, therefore, as anything that might be taken to be the 'purpose' of all the transmutations of nature. Can any one characteristic of the natural process be so described, making due allowance of course for the feeling of guilty reluctance we shall always feel in referring to a natural 'purpose' (see Chapter 1)?

The only such overall purpose, Monod believes, is the ampliation or expansion of DNA — a conception summed up by the schoolboy aphorism 'a chicken is merely the egg's way of making another egg'. The only reasonable ground on which one could object to this statement is the pejorative use of 'merely', for a chicken is a remarkable and breathtakingly ingenious way of making another egg, and among birds generally we find other ways of making eggs, always marvellously beautiful and ingenious. It is in such evidence as this — the marvel of the whole symphonic texture of the natural process — that the reverent should hope to find evidence of a Great Composer.

Unfortunately, the testimony of Design is only for those who, secure in their beliefs already, are in no need of confirmation. This is just as well, for there is no theological comfort in the ampliation of DNA and it is no use looking to evolution: the balance sheet of evolution has so closely written a debit column of all the blood and pain that goes into the natural process that not even the smoothest accountancy can make the transaction seem morally solvent according to any standards of morals that human beings are accustomed to.

Believers are no more likely to be shaken in their faith by the misgivings of geneticists than they were confirmed in them by the patronizing approbation of theoretical physicists; for faith rests upon quite other foundations — as secure to those who hold them as the derivation of a logical theorem.

Chapter Twenty-Four

The Great Amateur

People often wonder whether human beings are capable of further evolution. Leaving open the question of whether any such evolution will occur or not, the answer is assuredly 'Yes'. Human beings have a vast reservoir of inborn diversity and an open or 'wild type' breeding system which would make it possible for that diversity to be fully exploited; they have no extreme specialization such as the anteater's snout or the fly trap of an insectivorous plant — no specializations that commit them to one particular kind of life. Indeed, from an evolutionary point of view man is the great amateur among animals. A merely professional animal would probably have committed itself by structure or function to a bondage it could not now escape.

It is, however, very unlikely that any major evolutionary change will come about during the future life of man on earth; but although it is unlikely that any major evolutionary change will occur, small systemic changes in gene frequency — which are, in a technical sense, evolutionary — are quite likely to happen: the age of pandemics may not yet be over. Certain viruses which may hitherto have lived in comfortable symbiosis with man might become pathogenic in a mutant form, and if so the differences of genetic make-up could have a profound effect on our susceptibility, and our genetic make-up will change accordingly. There may also be a change in the tempo of ageing: as life extends farther and farther beyond the age of reproduction or as reproductive age becomes younger and younger, so the pressure of natural selection against harmful late-acting genes will progressively diminish and any senescent changes for which they are responsible will take even deeper root in the human population; this effect will become increasingly obtrusive because the chronological age at which people admit to getting on in years — e.g. to being middle aged — gets later and later.

Our reasons for thinking that no major evolutionary change will occur are twofold. In the first place the exercise of any artificial selection over very many generations would require acquiescence in the rule of a long dynasty of tyrants, and although such a tyranny is not inconceivable, such consistency of policy assuredly is. In the second place ordinary or endosomatic evolution (see Chapter 6) is no longer a principal agency for securing fitness in human populations.

A medical student once asked whether human beings might not evolve to possess wings and so make it possible to fly. The question was made particularly memorable by the fact that the student had to raise his voice to make it heard above the sound of a passing aircraft: a foolish question indeed, for it is obvious that human beings have already acquired some of the capabilities of both birds and fish — capabilities which they owe to their own special style of evolution, the 'exosomatic'.

Many important differences distinguish the long-term prospects of human beings from those of any other species of animal.

We have already explained (pp.11—12) that many fastidious biologists prefer to use the genteelism 'teleonomy', where Aristotle would have used 'teleology', in referring to the purposive or ostensibly goal-directed behaviour characteristic of living things.

With human beings no such niceties are called for. For good or ill, human behaviour is purposive: we do things because we intend to or fail in spite of our intentions. Human behaviour can be genuinely purposive because only human beings guide their behaviour by a knowledge of what happened before they were born and a preconception of what may happen after they are dead: thus only human beings find their way by a light that illumines more than the patch of ground they stand on.

The argument that human beings will not survive because most other animals have not is poor: the possibility of acting on good intentions differentiates us very sharply from other animals. To depreciate moral judgment on the ground that it has evolved because of its survival value is not a valid counter-argument: survival value is the very quality we are claiming for it.

At no time since the early years of the seventeenth century* have human thoughts been so darkened by an expectation of doom. In their apocalyptic moods people nowadays foresee a time when pressure of population will become insupportable, when greed and self-interest have so far despoiled the environment that the life of man will once again be solitary, poor, nasty, brutish and short, when international rivalry has brought commerce and communications to a standstill ...

> *When the great markets by the sea shut fast*
> *All that calm Sunday that goes on and on:*
> *When even lovers find their peace at last,*
> *And Earth is but a star, that once had shone.*
>   J.E. Flecker**

We, on the contrary, do not believe that any evil which may befall mankind cannot be prevented or that any evil which now besets it is irremediable. It is not reasonable on the one hand to cower at the havoc science and technology may cause in achieving the Baconian ambition of 'effecting all things possible' and at the same time to exclude from all things possible the discovery of remedies for technological male-factions. For remedies people look first to science and then look away in disappointment, partly because they mistake the nature of the problems and partly because they have grown so used to thinking of science and technology as a secular substitute for the miraculous; but most of the problems that beset mankind call for political, moral and administrative rather than scientific solutions.

---

* P.B. Medawar, *The Hope of Progress* (Wildwood House, London, 1974), pp.110–27.

** J.E. Flecker, *The Golden Journey to Samarkand* (London, 1913).

Another way in which human beings are amateurs in a professional world is that not all human activities have survival as their principal purpose. Even though our extra curricular activities are those that make life worth living — Mozart's piano sonatas and the paintings in the Uffizi Gallery amplify the human spirit and not human DNA (see Chapter 23) — nothing will reconvert human beings from amateurs into pros more quickly than the imminence of mortal danger. In this context, being professional may imply submitting again to the tyrannical philosophy of reproductive advantage that has brought us this long way already. Clearly some compromise between the amateur and the professional is called for.

Although it is widely regarded as frivolously superficial to suppose that the human predicament is remediable, nothing in reality could be more superficial than failure to realize that acquiescence in the notion of impending doom is a principal factor in helping it to come about. In spite of all its frightening groans and rattles, the great world machine can still be made to work, but not unless it comes to be accepted that the long-term welfare of human beings cannot be secured by policies that promote the interests of some people at the expense of others or even the interests of mankind at the expense of other living things. The *unity of nature* is not a slogan but a principle to the truth of which all natural processes bear witness. The lesson has been learnt too late to save some living creatures, but there may just be time to save the rest of us.

Glossary of Technical Terms

The Glossary explains more fully various technical terms that appear in the text. It also defines such confusing terms as 'secretion', which have a technical meaning widely different from their everyday usage. Terms in italics are included elsewhere in the Glossary.

ADAPTATION   (a) The process in which sense organs cease to arouse nerve impulses after long exposure to a uniform stimulus.   (b) Progressive change in a population of organisms that adapts its members better to their prevailing habit or environment.   (c) In bacteria, as (b) but specially referring to adaptation to use new foodstuffs or to resist the action of antibiotics.

ADAPTIVE RADIATION   An adaptive diversification, such that the several members of a group of organisms come to occupy and adapt themselves to a variety of different environments or habits of life.   Insects and marsupial mammals show high degrees of adaptive radiation.

ADENOVIRUS   A member of a large group of pathogenic viruses which cause a variety of diseases in domesticated animals and are incriminated in many infections of the upper respiratory tract in man.

ALLELE (= allelomorph)   In the earliest formulations of Mendelian theory it was believed that genetic determinants were always present in pairs and were responsible for the determination of alternative and contrasted characteristics (e.g. tall v short in pea plants or — as discovered very much later in human beings — tasting v non-tasting of the compound phenylthiourea). When this simple state of affairs is realized the alternative or 'allelic' genes occupy correspondingly opposite positions on each chromosome of a pair, so that in sexual reproduction allelic genes separate and enter one gamete or another, as chromosomes do.

AMINO-ACIDS   Amino (NH_2)-substitution products of a fatty acid: the monomer or building block of a *polypeptide*.

AMNION   The membrane enclosing a fluid-filled cavity which protects the embryos of reptiles, birds and mammals and provides them with, in effect, an aquatic environment.

ANTIBIOTIC   A product of one micro-organism that inhibits the growth of another, e.g. penicillin, streptomycin.

ANTIBODY   Soluble blood protein that reacts upon foreign substances in such a way as to destroy them or annul their action. See Chapter 13.

ANTISEPTIC SURGERY   A surgical technique in which antiseptics are used in an attempt to kill the bacteria that would otherwise give rise to wound infections. See *Aseptic Surgery*.

ASEPTIC SURGERY   A technique of surgery introduced by B.G.A. Moynihan of Leeds and W.S. Halsted of Johns Hopkins University in which wound infection is prevented by using sterile instruments, drapes, etc., and by operating with gloved hands.

BIOSPHERE   The domain or realm of living organisms.

BRAIN   The principal correlation centre and sensory input station in the nervous system. Divided in vertebrates into forebrain (prosencephalon), between-brain (diencephalon), mid-brain (mesencephalon), cerebellum and medulla oblongata. The cavities of the brain are referred to as ventricles.

BROWNIAN MOVEMENT   Random perturbation of small particles in fluid suspension particularly well shown by pollen grains in water and fat droplets in milk.

CELL   The *epigenetic* domain of a nucleus. Alternatively, the smallest sub-division of an organism capable of independent life.

CELL-MEDIATED IMMUNITY   See *Immunity.*

CHIMERA   Organism composed of cells descended from two different zygotes, *especially radiation chimera,* an organism in which blood-forming cells have been destroyed by radiation and replaced by blood-forming cells from a different organism.

CHROMOSOME   The vehicle of the nucleic acid that encodes genetical information. Chromosomes are present in pairs (the *diploid* condition) in the body cells of all sexually reproducing animals, but only one member of each pair is present in a gamete (the *haploid* condition).

CLONE   Cell lineage derived from a single cell by repeated cell divisions.

COLLOID   Solutions of very large molecules that are often asymmetric and bear electric charges behave differently from conventional solutions of crystalline solutes such as sugar or salts. These differences form the subject matter of 'colloid chemistry'. Blood *plasma,* lymph, cell-sap and most biological fluids are colloids. The 'phases' of a colloid are the number of physical states that enter into its make-up. Thus a polyphasic colloid, such as 'protoplasm' was thought to be, will consist of solid particles, fat droplets and liquid droplets in suspension in some continuous phase such as a salty liquid.

COMMENSALISM   The condition bordering on symbiosis in which organisms share food, though not necessarily with mutual benefit — e.g. small fish accompanying sharks that eat the scraps left from their meals.

COOLEY'S ANAEMIA   (= thalassaemia major)   The homozygous form of a congenital blood disorder prevalent in the Mediterranean basin and characterized by the reduced size of blood corpuscles and reduced quantity of haemoglobin in the blood.   The heterozygous state (thalassaemia minor), a lesser affliction, is widely reputed to confer some protection against malaria.

CORONARY ARTERIES   The principal arteries supplying the muscles of the heart.   So-called because they encircle the heart like a crown.

DIPLOID   See *Chromosome, Gametes.*

DOMINANCE   The state of affairs when one *allele* overrides the outward expression of another.   A dominant gene exercises the same outward effect when inherited from only one parent as when inherited from both.   Genes that make their effects apparent only when inherited from both parents are said to be *recessive.*

EMERGENCE   Philosophic doctrine opposed to *reducibility* which declares that in a *hierarchial* system each level may have properties and modes of behaviour peculiar to itself and not fully explicable by analytic reduction.

END ARTERY   An artery uniquely responsible for the blood supply of a particular tissue, e.g. the retinal artery.

END CELL   The final product of differentiation at the end of a sequence of cell divisions.   End cells do not divide and cannot therefore carry on a lineage.   Examples are nerve cells, red blood corpuscles and the cells of the horny layer of skin *epithelium.*

ENTELECHY   Complex   metaphysical   notion   of Aristotle's, referring to the endowment or state of affairs which brings it about that some potentiality is made actual. In a biological context Hans Driesch's entelechy was the animating or vital principle of living organisms.

EPIGENETIC   Having to do with the execution of genetical instructions.

EPITHELIUM (– A)   Contiguous cells of the same kind and the same orientation that bound a surface are referred to as epithelia (see p.129, Chapter 15).

ETHOLOGY   Science of animal behaviour, with special reference to natural behaviour and the identification of functionally meaningful patterns of activity.

FERTILIZATION   In sexually reproducing organisms, the union of sperm with an egg so combining the genetic information from male and female and activating the egg to start development.

FIBRIN   Insoluble protein that forms the fibres of a blood or lymph clot.

FITNESS   The enjoyment of selective advantage, or the property of conferring it. Organisms (or by extension, genes or genetic make-ups) that increase their representation in succeeding generations are said retrospectively to be fitter than those which do not.

GALACTOSEMIA   A serious inborn error of metabolism of recessive determination, marked by the presence of galactose in urine and high levels of galactose in the blood.

GAMETES   Reproductive cells – in males, sperm, and in females, eggs. Each contains the *haploid* number of chromosomes but on *fertilization* of the egg the *diploid* number is restored.

GENE   The least element of the *genome* that may be held responsible for an inherited character-difference.

GENOME   Genetic apparatus of an organism considered as a whole and as characteristic of it, e.g. the 'human genome', referring to the chromosomal make-up characteristic of human beings.

GENOTYPE   Genetic constitution of an organism as opposed to its overt character make-up or *phenotype*.

GONADOTROPHINS   Pituitary hormones that stimulate the development of the gonads and the maturation of the gametes.

HAEMOGLOBIN   Red oxygen-carrying pigment characteristic of the blood of vertebrates, but also found in some lower animals.   In carrying oxygen haemoglobin is not 'oxidized' in the chemical sense, but forms an easily reversible attachment to oxygen referred to as 'oxygenation'.

HAEMOPHILIA   A congenital aberration of blood clotting mechanism leading to excessive bleeding.   A sex-linked form is manifested by males but transmitted through females.

HAPLOID   See *Chromosome, Gametes.*

HARDY-WEINBERG THEOREM   An algebraic rule for translating statements about gene-frequency into statements about the frequency of *genotypes* in a population.

HETEROZYGOTE   An organism is described as 'heterozygous' for a specified gene pair when it has inherited *unlike alleles* from its two parents.   If the two alleles are the same, the organism is *homozygous* with respect to that locus.   The corresponding substantives are 'heterozygote' and 'homozygote'.   Outbreeding organisms are heterozygous for most gene pairs and homozygosity for most gene *loci* is a rarity which can only be approximated to by prolonged inbreeding.

HIERARCHY   A system of ordering typified by an Army command, in which a complex structural organization can be resolved into levels or grades, each subordinate to the one above it, but each having a comparable formal structure. In biology the system/society/organism/organ/cell is such a hierarchial system.

HOLISM   The metaphysical doctrine that all living systems tend to form highly integrated and indivisible entities.

HOMOLOGY   The correspondence of structure or of any biological performance attributable to correspondence of developmental processes or of genetic determination, e.g. homology of birds' wings with mammalian forelegs.

HOMOZYGOTE   See *Heterozygote*.

HORMONE   Chemical agent circulating in bloodstream by which an event in one part of the body influences or initiates an event in another part.

HUMORAL IMMUNITY   Immunity mediated through soluble bloodborne substances, especially *antibodies: cell-mediated immunity* (C.M.I) mediated through circulating cells of lymphoid family. See *Lymphocyte*.

HYBRID VIGOUR (= heterosis)   An accession of physical vigour and resistance to disease, often accompanied by increased size and growth rate, associated with the hybridization of animals which, because of *inbreeding* or maintenance as 'breeds', have become *homozygous* for many gene *loci*.

IMMUNITY   Resistant or refractory state arising from an adaptive response of an organism to a non-Self substance.

INBREEDING   Systematic propagation of a lineage through genetically related organisms, reaching its extreme form in self-fertilization, where it is possible.   Free-living organisms are characteristically outbreeders.   Breeding tends to *homozygosity* where outbreeding favours *heterozygosity*.

INFORMATION   Stands for order or orderliness as it may be embodied in a material structure or in a message specifying its assembly, as exemplified by the genetic signal embodied in chromosomes.   The negative of information is disorder or, in a thermodynamic context, entropy. Obtrusive signals conveying no information and diminishing order are dismissed as noise.

ISOTOPES   Isotopes are variant forms of chemical elements that differ in atomic weight or radioactivity from their more abundant forms. They have the same chemical properties as the corresponding normal forms, and are treated alike by living organisms. Examples of isotopes are the heavy hydrogens deuterium and tritium. Deuterium forms the 'heavy water' (D_2O). Tritium is radioactive, emitting beta particles — a property which makes it specially valuable as a molecular label in biological research. A compound in which ^3H substitutes for hydrogen is described as 'tritiated'.

KLINEFELTER'S SYNDROME   A disorder of sexual development associated with chromosomal abnormality; gonads develop imperfectly but *gonadotrophin* levels are high and there may be excessive development of the mammary glands.

LOCUS, LOCI   Position(s) on the chromosome occupied by allelic genes. The locus occupied by one allele corresponds anatomically to the locus occupied by the other on the second chromosome of the pair.

LUDDISM   About 1810 Ned Ludd and bands of hand-workers destroyed the textile machinery they feared would put them out of work. Luddism thus came to be a general name for opposition to technological advance and has been used in a medical context to describe a modern movement that opposes the use of the mechanized life-saving apparatus, characteristic of intensive care.

LYMPH   See *Lymphatics*.

LYMPHATICS   The vessels that drain the tissues and after passing through lymph nodes open into the venous system. The content of lymphatics is a colourless clotting fluid, *lymph*.

LYMPHOCYTES   Circulating white blood corpuscles of two families: B lymphocytes, concerned with *antibody*-formation, and T lymphocytes, agents of *cell-mediated immunity*.

MALTHUSIAN PARAMETER   One of many proposed single-value measures of reproductive capacity of a population. It represents the rate of compound interest at which the population could increase or decrease if the prevailing fertility and mortality continued over many generations. The value may be positive or negative. See also *Net Reproduction Ratio*.

MAPPING (From Cartography)   Transformation of one form of order into another, normally in such a way as to establish a continuous one-to-one correlation between the new and the old order.

METABOLISM   General name for the chemical transformations that occur in nutrition, growth, energy storage, energy expenditure and all physiological performances of the body.

METAMORPHOSIS   The radical internal and external transformation that accompanies development in organisms whose embryos differ greatly in structure or habit from their corresponding adult forms, e.g. metamorphosis of tadpole into frog, or of caterpillar into chrysalis and then into butterfly or moth.

MITOCHONDRION (– A)   Cellular organelle of biogenetic origin regarded as the seat of cellular respiration. It is unusual in its possession of extra-nuclear DNA.

MUTAGEN   An agent that increases the frequency of *mutation*, e.g. X-irradiation.

MUTATION   The rare molecular accident by which new genetic information comes into being. A gene which has undergone mutation is described as a 'mutant'. So also is the organism in which the mutant gene expresses itself.

MYCOPLASMA   Although the prefix *myco* usually signifies something to do with a fungus, mycoplasmas are usually thought of as minute, naked bacteria, i.e. as bacteria lacking their complex cell walls. Mycoplasmas are widely distributed throughout the respiratory, genital and digestive tracts of domesticated animals including man. Also known as PPLO (= pleuropneumonia-like organisms).

NET REPRODUCTION RATIO   One of several proposed single-value measures of reproductive capacity of a population. May be taken to represent ratio of live female births in successive generations if prevailing birth rates and death rates continue over many generations. Its value may be more or less than unity, which represents net replacement. See *Malthusian Parameter*.

OSMOTIC PRESSURE   The pressure (sometimes very great) that is generated by the diffusion of water from a region of higher to a region of lower concentration – e.g. the diffusion of pure water into a solution across a membrane which does not allow the dissolved substance to pass through.

PARASITE   Organism that lives at the expense of another, either attached to its surface (ectoparasite) or living internally (endoparasite). *Obligative* parasites cannot live apart from organisms they parasitize (their 'hosts'). In their several ways cuckoos, tapeworms, many eelworms and some fungi are parasitic. Mutual parasitism is *symbiosis*.

PHAGOCYTOSIS   Eating by engulfment of solid particles characteristic of cells of amoeboid type.

PHASE CONTRAST MICROSCOPE   A microscope using ordinary light of visible wavelengths but specially well adapted to studying living cells. The objects inspected are distinguished not merely by refractility but also by the degree to which they alter the phase of transmitted light.

PHENOTYPE   See *Genotype*.

PHENYLKETONURIA   An inborn error of the *metabolism* of the *amino acid* phenylkalanine, leading to severe mental retardation; *recessive* genetic determination.

PINOCYTOSIS   Imbibition of minute droplets of fluid into the cell.

PLANKTON   Minute animals and plants of the surface waters of the ocean whose own powers of horizontal movement are negligible compared to ocean currents and other great movements of water.

PLASMA   The fluid portion of blood, i.e. blood minus corpuscles.

POETISM   The acceptance of a hypothesis from literary rather than scientific inducements.

POLYMER   A complex molecule formed by the compounding of smaller building elements, monomers, of the same general structure; thus, polysaccharides are generally polymers of 6-carbon sugars and proteins of *amino acids*.

POLYMORPHISM   The stable sub-division of a population into genetically distinct types, as of human blood groups.

POLYPEPTIDE   The backbone of the protein molecule. A *polymer* of amino acids.

POLYPHASIC   See *Colloid.*

PSITTACOSIS   A virus disease of birds, transmissible (especially by parrots) to human beings, in whom it causes a febrile illness with symptoms not unlike those of typhoid fever.

RADIATION CHIMERA   See *Chimera.*

RECESSIVE   A gene that exercises little or no outward effect unless it is present in both chromosomes of a pair, and therefore has been inherited from both parents, is said to be recessive. Contrast *Dominance.*

REDUCIBILITY   Philosophic doctrine that in a *hierarchy* all the properties of one level or grade of the system can be explained in terms of the properties of a lesser or subordinate grade in the system. Thus, according to this canon, there is nothing about organisms that cannot be explained by the study of cells or about cells that cannot be explained by the study of molecules.

RUMEN One of the digestive chambers of animals such as the cow that live largely on foods very rich in cellulose, i.e. the polymers that form plant cell walls. The rumen contains a huge population of protozoa and bacteria which break down protein, fat and cellulose. Rumination is a regurgitation of food from the rumen into the mouth where it is finely communited ('chewing the cud') before passing finally into the digestive tract.

SECRETION Literally the separation of one biological substance from another, usually followed by its discharge. In this sense tear glands secrete tears and sweat glands sweat, but 'secretion' has come to refer to the elaboration and discharge of any substance synthesized in a glandular cell.

SERUM The fluid portion of blood after coagulation, i.e. approximately *plasma* minus fibrin.

SOMA Ordinary body tissue as distinct from the reproductive cells or *gametes*. In Weismann's terminology soma was distinguished from 'germ-plasm'.

SPECIFICITY The exactly complementary relationship between an agent and something acted on, or between instruction and performance fulfilled.

STEROIDS The family of chemicals typified by cholesterol, to which belong the sex hormones, cortisone and Vitamin D. Sterols are typically waxy substances, insoluble in water.

SYMBIOSIS Living together in a state that brings mutual benefit to the symbionts participating, e.g. symbiosis between fungus and a simple form of green plant to form a lichen. See also *Commensalism.*

SYNGENEIC   Especially in the context of transplantation, having as nearly as possible the same genetic make-up.

TAXONOMY   Science or art of arrangement or classifying, particularly of living organisms.

TELEOLOGY   The science of purposes or 'final causes', or the metaphysical doctrine that all bodily structures and performances are determined by purposes they fulfil.

TELEONOMY   The neutral term (contrast *Teleology*) describing but not explaining the quasi purposive or goal-directed activities of living things.

THYMUS   A lymphoid organ very prominent in young warm-blooded animals, and responsible for the maturation of the *lymphocytes* that transact *cell-mediated immunity.* It is normally reduced in size in adult life.

TRACHOMA   A chronic virus infection causing an inflammation of the conjunctiva which, in severe cases, leads to the penetration of the cornea by blood vessels.

TRANSDUCER   An instrument that converts one wave form or energy form into another.

TURNER'S SYNDROME   An abnormality of sexual development associated with chromosomal disorder. Marked by impaired development of the ovary, a reduction of height and a variety of skeletal and dental defects.

UREA   Soluble nitrogen-containing compound present in urine and representing a late stage in the breakdown of proteins. Urea is widely used as a fertilizer and is the monomer of urea-formaldehyde resins.

VIRUS   An infective nucleic acid (RNA and DNA) that

subverts the synthetic machinery of living cells in such a way that more copies of itself are produced. In the infective form of the virus the nucleic acid is normally wrapped in a protein capsule.

ZYGOTE   The *diploid* cell resulting from the union between sperm and egg, and the starting point of the development of all sexually-reproducing organisms.

Index